Experiments in Logic and Computer Design

AF718993

WITHDRAWN

Experiments in Logic and Computer Design

Albert Y. Teng

AT&T Bell Laboratories

William A. Malmgren, Sr.

GTE Business Communications Systems

PRENTICE-HALL, INC., *Englewood Cliffs, NJ 07632*

Library of Congress Cataloging in Publication Data

Teng, Albert Y.
Experiments in logic and computer design.

Bibliography: p.
Includes index.
1. Logic circuits—Experiments. 2. Computers—Circuits—Experiments. I. Malmgren, William A. II. Title.
TK7868.L6T45 1984 621.3819'582 84-1935
ISBN 0-13-295833-3

Editorial production supervision: *Barbara A. Cassel*
Manufacturing buyer: *Anthony Caruso*

TK
7868
.L6
T45
1984

©1984 by AT&T Bell Laboratories

All rights reserved. No part of this book may be reproduced, in any form or by any means, without permission in writing from the publisher and AT&T Bell Laboratories.

Printed in the United States of America

10 9 8 7 6 5 4 3 2 1

ISBN 0-13-295833-3 01

Prentice-Hall International, Inc., *London*
Prentice-Hall of Australia Pty. Limited, *Sydney*
Editora Prentice-Hall do Brasil, Ltda., *Rio de Janeiro*
Prentice-Hall Canada Inc., *Toronto*
Prentice-Hall of India Private Limited, *New Delhi*
Prentice-Hall of Japan, Inc., *Tokyo*
Prentice-Hall of Southeast Asia Pte. Ltd., *Singapore*
Whitehall Books Limited, *Wellington, New Zealand*

Contents

Preface

By the billions, Integrated Circuits (ICs) are moving into schools, homes, and offices. To succeed in the future, everyone will need to know something about the functions and applications of these important IC chips. We believe that it is possible to package the important new IC developments into a series of logic design experiments that can educate beginners and give them an understanding of this new technology. Every effort has been made to ensure that the material in this book emphasizes the important computer hardware concepts and applications.

Many experimental books have been published on the subject of elementary logic design or specific vendors' microprocessors. The purpose of this book is to provide hands-on experience in general digital logic and computer systems through a set of carefully developed experiments. It is intended to serve a large class of computer scientists, engineers, managers, technicians, and anyone who wants to gain an insight into the complexities of IC applications. The book assumes that the reader has little or no electronic hardware background. The only prerequisites are some experience in assembly language programming and a basic knowledge of boolean algebra and digital systems. After performing the experiments, the reader should be able to design sophisticated digital systems using off-the-shelf IC devices.

Originally, the experiments in this book were designed for a graduate or undergraduate course in computer architecture. The drafted version of the book has been used at the Computer Science Department, Illinois Institute of Technology, since the Spring of 1980, to augment a computer architecture textbook. The experiments could be performed by using any one of the commercially available digital trainers, or by breadboarding the individual components using available equipment and materials. Our objective is to design an inexpensive and useful series of experiments for introductory digital system or computer architecture courses. The experiments can be used for a companion laboratory to many existing logic design textbooks.

Chapter 1 is an introduction to IC technology. We introduce the material and equipment in Chapter 2 and IC circuit construction techniques in Chapter 3 before the discussion of experiments. Chapter 4 has six simple experiments for instructional purposes. The design is completely described for each experiment; all you need to do is to implement the circuit. Chapter 5 has six small digital system experiments that are more complicated and require paper-and-pencil design before implementing the circuitry on the breadboard. The design experiments will provide you with some hands-on design and implementation experience by using the circuit construction skill learned from the instructional experiments. You can actually propose your own design of sophisticated digital functions and then verify its correctness by implementation.

Chapter 6 contains three digital design projects that allow readers to design their own circuits using only general design requirements and suggestions. We believe that the design experience gained through the completion of these projects will be the most beneficial part of this book. These design projects may be too complicated to be implemented by using the IC chips suggested in the beginning of each experiment. The purpose of these projects is to illustrate some useful hardware applications requiring sophisticated design processes. The computer in the design project does not support any complex programming, but it works as a stand-alone free-running device. The sorting engine illustrates an example of implementing software functions by VLSI devices. The computer network interface design covers a wide range of topics, including local area network design, data communication protocols, and network performance analysis. The bibliography and the appendixes appear at the end of the book.

The senior author wishes to thank his architecture students who helped him in this endeavor. Their keen interest in computer architecture has made the painstaking preparation of this book worthwhile. The authors especially would like to thank John Heck and Suh-Fen Teng of AT&T Bell Laboratories for their invaluable suggestions and for reviewing many versions of the manuscript. John has also contributed by designing the wiring schematic diagrams of IC chips.

We also wish to thank Anthony Wojcik, Martha Evens, Karen Coates, and Harold Jackson, who gave us a great deal of encouragement and support during the development of the project. We deeply appreciate the generous support of the Text Processing, Art, and Technical Publication Departments of AT&T Bell Laboratories. The insightful critique by Prentice-Hall reviewers was also very helpful.

Naperville, Illinois

Albert Yanpyng Teng
William A. Malmgren, Sr.

Experiments in Logic and Computer Design

1

Introduction

The electronic revolution began in 1947 when three Bell Laboratories scientists invented the transistor (short for transfer resistance) which was faster, smaller, consumed less energy, and had fewer failures than vacuum-tube technology. The transistor was the foundation of a new technology: Integrated Circuits (ICs). In 1959, scientists at Texas Instruments and Fairchild Camera and Instrument Company introduced ICs - a single semiconductor chip with Small Scale Integration (SSI) of one to ten gates. By 1965, several laboratories were producing chips with Medium Scale Integration (MSI) of 10 to 100 gates. In 1970, Large Scale Integration (LSI) of thousands of gates was introduced. Later, in 1971, Intel, a microelectronics company, put an entire computer circuit on a chip and the IC became a microprocessor. Today, with the advent of Very Large Scale Integration (VLSI) it is possible to fabricate chips containing close to one million components per chip. WE* 32000 microprocessor contains nearly 600,000 components and can support the UNIX™ operating system and the C language.

The density of IC chips has approximately doubled every year since 1959. The closer the electronic components are to each other, the faster the circuit can perform. High density ICs have small size and

---

* Trademark of AT&T Technologies.

weight, are reliable, and consume a small amount of power. Another key advantage of IC technology is the ability to produce numerous chips at one time so as to cut down the unit cost. The price of IC chips has decreased about 25 percent every year since 1959. Today, a tiny chip no larger than a postage stamp may be more powerful than a wall-to-wall computer of the early 1960s. The new IC technology is so fast that its actions are measured in picoseconds (trillionths of a second). Low cost, combined with high reliability and processing power, have pushed the application of IC chips in all fields. According to estimates, 2 million IC-based home computers were sold in 1982, and sales are expected to increase rapidly in the future.

The electronic industry has experienced a fast and extensive development since the birth of the IC. The IC can perform very simple or very sophisticated analog and digital functions with good performance, efficiency, and reliability. It can help children learn to talk and spell, control home heating and lighting systems, enable an automobile to talk, play bridge and chess, provide an entire "computer on a chip" (microprocessor), and propel us into the age of robots and information. Almost every appliance is likely to use IC chips in the near future. Over the next 5 years, IC chip sales are expected to grow by at least 20 percent annually, and the microprocessors will grow by 50 percent each year. Many scientists predict the IC-based computer will be the single most significant technological force in existence by the end of the century. Scientists call ICs one of the greatest technological achievements of mankind and predict ICs will create profound changes in society similar to the automobile in the early part of the 20th century.

An increasing amount of software has been implemented in silicon. A single-chip 32-bit microprocessor, WE 32000, introduced by AT&T Bell Laboratories, has hardware support for sophisticated operating system functions such as process context switching and system calls. Motorola supplies a "real-time operating system" within the 6801 chip itself. National Bureau of Standard's "Data Encryption Standard" is another example of IC implementation, in this case, for data security measures. It is also likely to have a new communication chip which uses built-in instructions to replace thousands of lines of software in performing complete logical control of a data link according to the X.25 protocol for public, packet-switched data networks. Software support by silicon chips is the direction for people who want to design systems with real-time performance. We will see more operating system, communication, and application software in hardware chips in the future.

The drastic growth in numbers and applications of IC-based digital systems has brought a new challenge to computer software and

Applications of IC-Based Computers

applications engineers who are struggling to stay abreast of the state of the art. This book is written for persons with little or no hardware background. Every effort is made to ensure that the focus is on the important applications of ICs rather than the details of hardware theories. The ICs are viewed simply as "black boxes" that will perform desired functions when power voltage levels, corresponding to logic states, are presented at the inputs. The useful basis for designing experiments in this book is the knowledge of combinational circuits, analysis and synthesis of sequential circuits, and the Register Transfer (RT) language descriptions of hardware circuits. These topics are covered in the Appendixes.

The building-block concept in hardware design is similar to the module concept in software design. A module or a building block hides common functions that are likely to be changed or shared by other modules in order to minimize the effect of the future change on the rest of the system. We can combine the basic building blocks into a hardware board of chips for the ease of sharing and maintenance change. We can also combine the software modules into a subsystem for the ease of software sharing, maintenance, and change. The digital functions such as NAND gates, JK Flip Flop (FF) and Random Access Memory (RAM) may be packaged into different IC chips. The chips may be grouped together in a hardware board to provide a higher level function. The hardware boards can then be used in various electronic appliances, peripheral devices, or computers for the actual application. Similarly, the experiments in this book require you to select the IC chips for the required digital functions so as to implement a more sophisticated prototype circuitry on a testing breadboard.

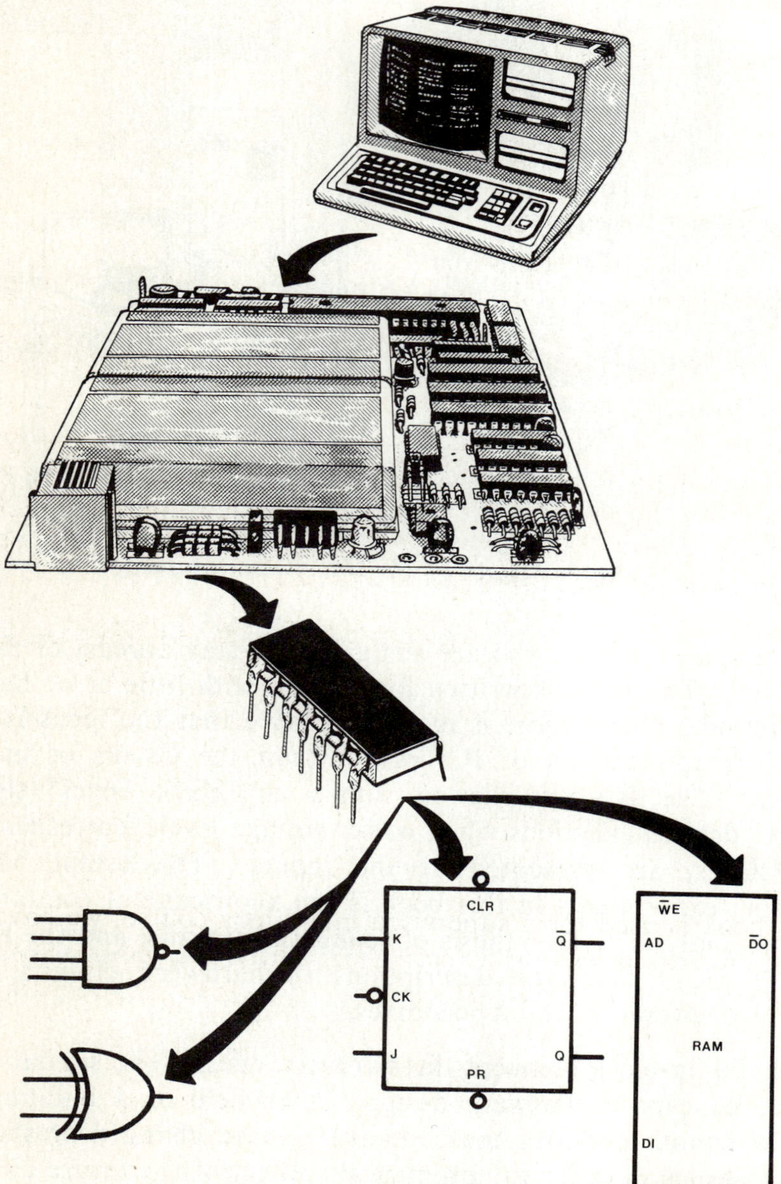

Hierarchy of Building Blocks

You should more fully appreciate the building-block approach used in the experiments in this book at the end of the study. The simple material and circuits are presented first. More advanced subjects and functions are introduced later by using some of the earlier results. A systematic scheme will also be introduced to illustrate a top-down design for the more complicated digital circuits in this book.

The experiments and projects in this book are constructed from basic building blocks such as registers, counters, adders, comparators, buses, and memories. You will study the necessary circuitry of the building blocks in the form of ICs. It is very important to be familiar with the various modules commonly encountered in IC packages. More than ten different ICs are introduced throughout the book.

The chips used in this book are all SSI and MSI devices. These devices are used as the basic units of building blocks for constructing experimental circuits. Typical SSI chips are basic logic gates and FFs. Typical MSI chips are counters, registers, and small memories. Typical LSI chips are memories, microprocessors, digital clocks, and calculators. Typical VLSI chips are 32-bit microcomputers, large-capacity sorting machines or memories, and sophisticated communication components. The experiments using sophisticated LSI and VLSI chips are more expensive to implement and are beyond the scope of this book. After completing the experiments in this book, you should, however, be able to apply the same techniques to designs using LSI and VLSI devices. (See VLSI Systems References in the Bibliography.)

The basic parts kit required for this book is listed in Chapter 2. If you can afford to use larger ROMs and RAMs, the AM 2900 series bit-slices provide an attractive way to permit a more substantial microcomputer project in Chapter 6. The availability of larger ROMs and RAMs can also support the experimentation on a more substantial microprogramming project similar to the programmed digital controller in Experiment 12.

If this book is used for a supervised laboratory course, we have a few suggestions based on our experience. There are more experiments than needed for a 1-semester course in this book and a subset of them should be selected. The time allowed to implement an experiment is recommended in the beginning of each chapter. All of the instructional experiments should be implemented by individuals who do not have any previous hardware experience. The selected design experiments should be implemented by a group of two team members. The design experiment is more complicated and may need extra equipment to implement. The design projects are recommended to be studied individually without the actual wiring of the circuit, due to the complexity of the circuits.

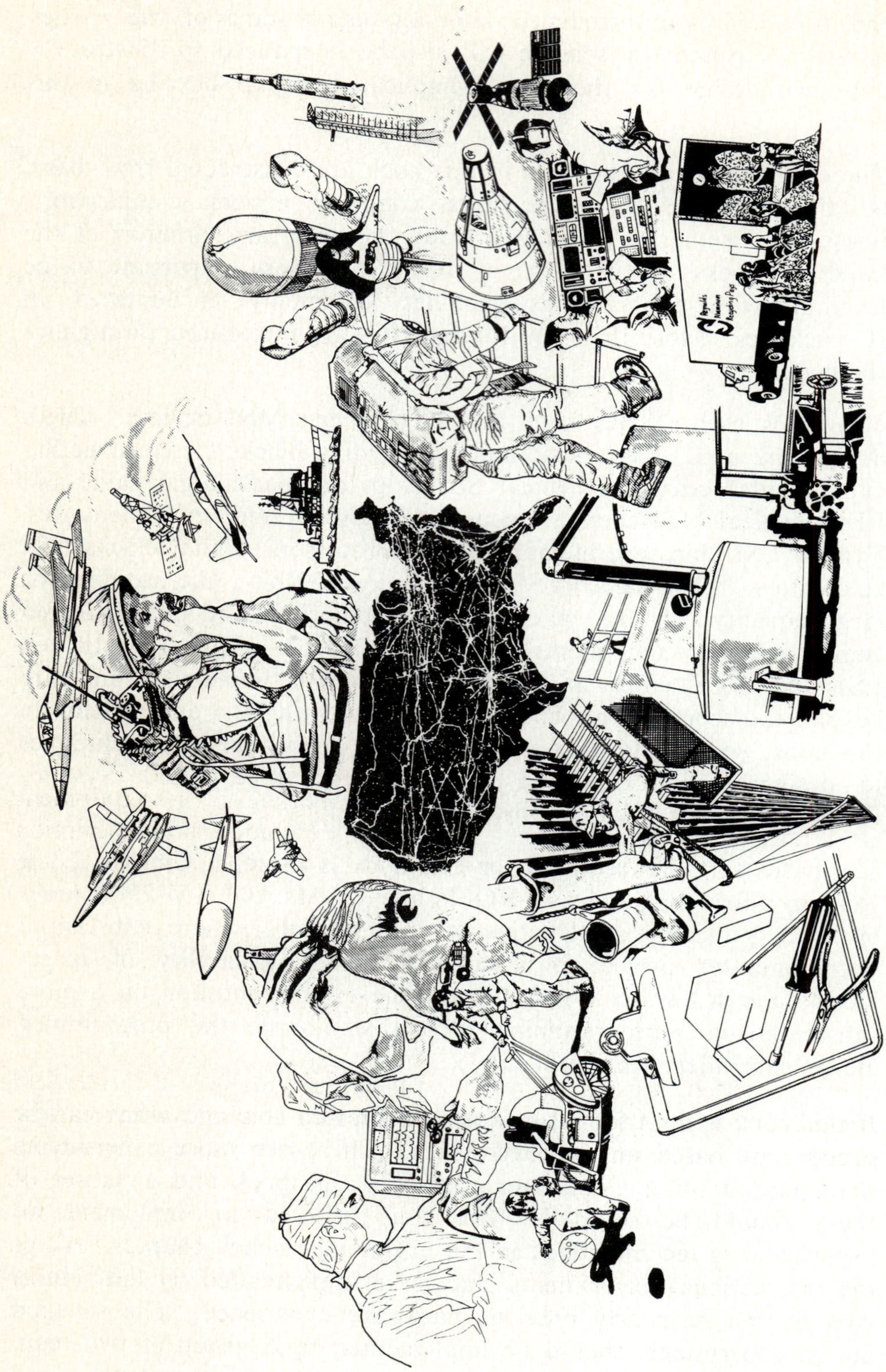

Applications of Integrated Circuits

2

Equipment and Material Description

This chapter gives a description of the material and equipment needed to do the experiments in this book. Included is a description of Logic Trainers and a description of the 7400-series of ICs. (See Reference 3 of Logic Design in the Bibliography.)

2.1 Equipment

The experiments in the book can be constructed by using any of the commercially available trainers or by breadboarding the desired components. If you do not use a trainer, you should use a solderless breadboard with strips of contacts into which ICs and wires can be easily inserted and removed. A 5-volt power supply must be available and the breadboard should accommodate at least seven 14-pin chips for the instructional experiments. The Heathkit ET-3200B Logic Trainer and ET-3300B Extended Breadboard are described here because they have been used to construct all of the experiments in this book; however, you can build or buy any other compatible trainer.

2.1.1 Logic Trainer

The Heathkit ET-3200B Logic Trainer, shown in Figure 2-1, contains the following items:

- four LED indicators,
- three power supplies (+12, -12, and +5 volts) and a ground,
- a clock,
- two logic switches which can be used to generate single pulses,
- four data switches having the up position logic 1, and
- a breadboard that can accommodate up to seven ICs.

The data switches allow you to input various combinations of digital variables and the LED indicators monitor the system responses. For those experiments requiring extra switches or indicators, outboard switches and LEDs can be used. Extra indicators and switches help debug complex circuits and determine how the circuit is functioning. The 1-Hz clock is usually the only clock pulse needed since the experiments in this book are operated under slow speed for easy observation.

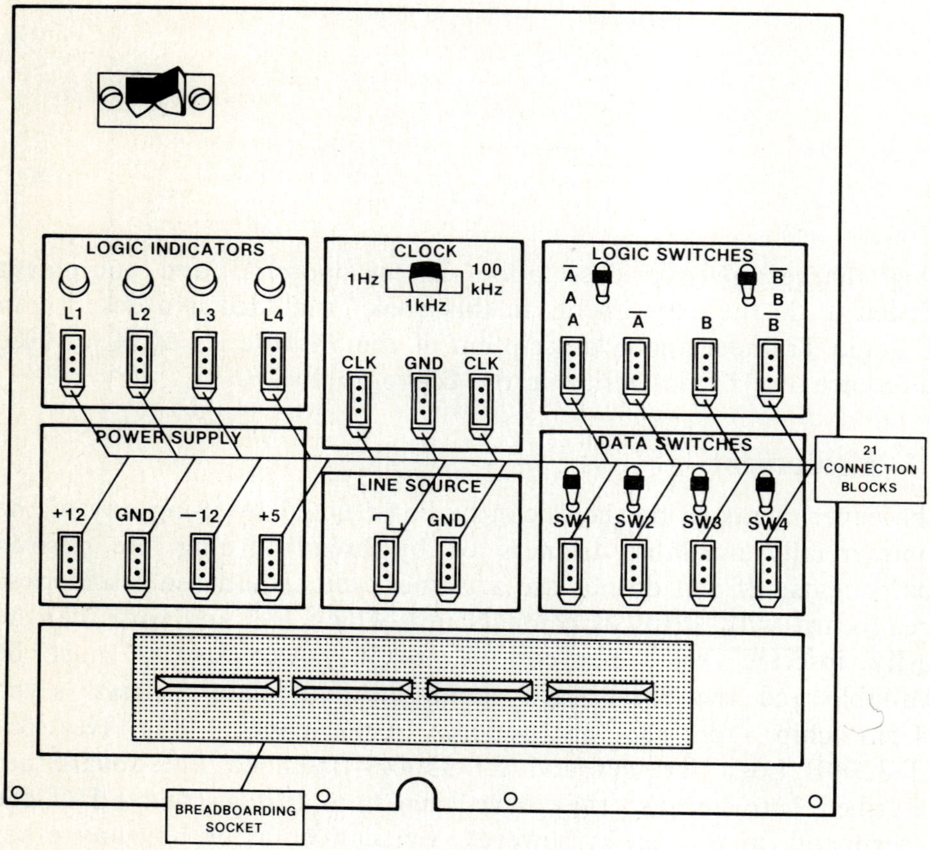

Figure 2-1. *ET-3200B Heathkit Logic Trainer (Courtesy of Heathkit/Zenith Educational Systems, a Division of Heath Company)*

2.1.2 Extended Breadboard

The Heathkit ET-3300B Extended Breadboard can be added to the logic trainer to accommodate more ICs for the more complicated design experiments. If you use the extended breadboard, run a ground wire from one of the logic trainer grounds to one of the extended breadboard grounds. The extended breadboard, shown in Figure 2-2, contains the following items:

- three power supplies (+12, -12, and +5 volts) and a ground, and
- four breadboards, each holding up to five ICs.

The +12 volt and -12 volt power supplies for both trainers are required for MOS IC chips. Experiments in this book use TTL chips that require only a +5 volt power supply and a ground connection.

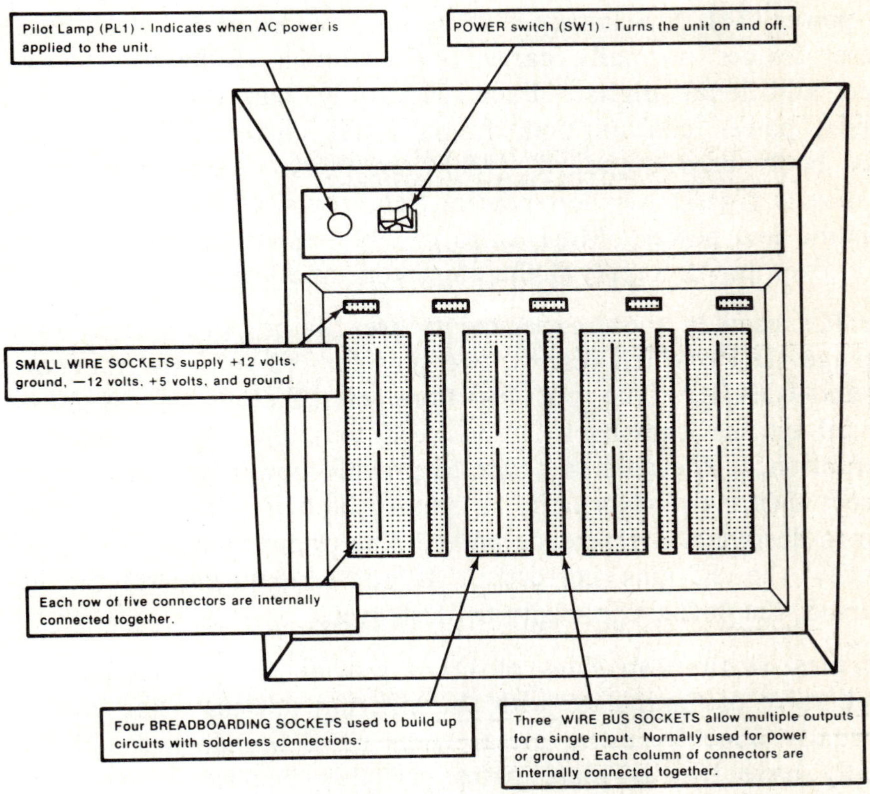

Figure 2-2. *ET-3300B Heathkit Extended Breadboard (Courtesy of Heathkit/Zenith Educational Systems, a Division of Heath Company)*

2.2 Materials

The materials required for constructing the experiments in this book are described in the rest of the section.

2.2.1 Integrated Circuits

Depending on the fabrication technologies, IC chips may be classified as Emitter Coupled Logic (ECL), Integrated Injection Logic (IIL), Transistor/Transistor Logic (TTL), P Channel Metal Oxide Semiconductor (PMOS), Complementary Metal Oxide Semiconductor (CMOS), or MOS devices. The fabrication procedures determine the critical parameters of IC such as speed, power dissipation, packing density, drive capabilities, and operating voltage levels. Although the members of one particular fabrication family are fully compatible with each other, it is generally difficult to match members of different families. All ICs used in this book are 74-series TTL family chips.

The 74-series TTL family can be subdivided into several distinct groups. Five major groups are standard (74-), high-power (74H-), low-power (74L-), Schottky (74S-), and low-power Schottky (74LS-). These devices normally carry a 74- prefix, followed by two or three additional digits. Each subfamily has different switching speeds, power consumption, fanout, and current drive capability. This book uses standard (74-) notations. You may also use low-power (74L-) or low-power Schottky (74LS-) devices which consume less power. More chips can be supported on a breadboard for a given power supply if the chips consume less power.

The ICs used in these experiments are 14- or 16-pin Dual In Line Package (DIP) which reserve two pins for +5 volt power (Vcc) and Ground (GND). The pin functions of IC chips are usually not printed on the chip itself. You need to know how to identify the orientation of the chip in order to provide power and ground to the correct terminals. Normally, a small notch is carved out from the chip to designate a reference point. The orientation of IC is vital to identify the IC pins correctly. If the orientation of the IC is reversed, the device can easily be destroyed.

The pins at the two sides of DIPs provide the connection for IC elements to communicate with the external devices. The connection is usually done by using metal wires on a breadboard. SSI chips usually have two or more gates per chip because it costs less to produce two gates on one chip than to have one gate on each of two chips. Gates are packaged in pairs (dual), in groups of three (triple), in groups of four (quad), or in groups of six (hex).

The manufacturer's data books are the best sources of information on ICs. The detailed description of IC chips in a manufacturer's data book usually contains the function table and the assignment of pins to functions. It also contains the necessary electrical characteristics and specifications needed by logic designers. The electrical specifications part is omitted in this book since it is not required for *learning* the logic design. This book emphasizes the design aspects rather than the hardware details.

2.2.2 Parts Kit

The parts kit, shown in Table 2-1, lists the type and quantity of components needed to construct all the experiments in this book.

Table 2-1. *Parts Kit*

Description	*TTL IC No.*	*Quantity*	*Fig. No.*
Quad 2-Input NAND Gates	7400	10	D-1
Dual 4-Input NAND Gates	7420	4	D-2
4-Bit Bistable Latches	7475	1	D-3
Dual JK FFs	7476	6	D-4
Quad 2-Input XOR Gates	7486	1	D-5
16x4 RAM Memory	7489	1	D-6
4x16 Decoder	74154	1	D-7
7-Segment Display Driver	7448	1	D-8
10K-Ohm Resistors	---	4	4-8
150-Ohm Resistors	---	4	4-8
7-Segment Display	HP5082-7740	1	-
LED(red)	XC556R	4	4-8
22-Gauge Solid Wire (Four colors)	-	4 feet (each)	-
DIP switch	206-7(SPST)	Optional	-
Hex Inverter	7404	Optional	D-9
Quad 2-Input AND	7408	Optional	D-10
Quad 2-Input OR	7432	Optional	D-11
4-Bit Binary Counter	7493	Optional	D-12
4-Bit Shift Register	7495	Optional	D-13

In addition to the parts kit shown in Table 2-1, you will need a few hand tools, such as needle-nose pliers and a wire cutter/stripper. An IC insertion tool and pin straightener are handy to have, but are not required.

Logic diagrams and function tables for all TTL ICs listed in the parts kit are illustrated in Figures D-1 through D-13 (see Reference 1 of Logic Design in the Bibliography). The optional parts listed in the parts kit are used to implement some optional experiment steps (such as the shift register and the counter) or to simplify the wiring connections (for example, using DIP switches instead of self-made switches, or using AND and OR gates instead of NAND gates). The required and optional ICs of the parts kit are adequate for constructing the experiments in this book. However, you may want to purchase some extra MSI chips to reduce the number of chips needed for the implementation of the more complicated experiments (for example, using larger RAMs for Project 1 in Chapter 6). The parts may be ordered from an IC supplier or catalog store.

As an example of reading one of the IC layout figures, let us consider the 7400 chip in Figure D-1. It has four 2-input NAND gates in one chip. Fourteen IC pins are numbered as indicated in the diagram. The 7400 pin locations for the power (Vcc) and ground (GND) are typical for the chips used in this book but are not standard. The outputs (1Y; 2Y; 3Y; and 4Y) are the NAND function of the inputs (1A, 1B; 2A, 2B; 3A, 3B; and 4A, 4B). Most of the 74-series chip circuits are quite complicated, and function tables are used to describe their operations.

3

IC Circuit Construction Techniques

After you have designed a circuit, you can verify the operation of the design by breadboarding. Breadboarding is the act of building a prototype model in a form that makes it convenient to make changes and corrections in the circuit. The actual circuit wiring is facilitated with a circuit connecting tool, called a breadboard or superstrip. No soldering is required to interconnect wires between ICs by breadboarding. Wires of designated lengths are cut and stripped at either end. Each end of the wire is plugged into one of the appropriate holes in the breadboard associated with a specific IC pin.

The wires should lie flat on the surface of the breadboard. The center gutter is used for inserting 14-, 16-, or 24-pin IC chips. Number *22-gauge solid* wire (not stranded) is usually used for connections to the breadboards. It is important not to use a wire gauge thicker than the size required by the specific breadboard used, since the contacts on the breadboard may be destroyed. Similarly, using a wire gauge thinner than the desired gauge may not ensure a sound connection and unexplainable circuit behavior may result.

You should turn off the power before you insert or remove an IC. Use long needle-nose pliers for grasping wires close to where they will be inserted into the breadboard. This prevents the wire from buckling. Insert a chip by first aligning the pins with the holes in the strip and then pressing it uniformly into the strip. You may have to adjust the distance between the two rows of pins before the insertion. To remove ICs from the breadboard, slide a screwdriver or an IC chip extractor under the body of the chip in the center gutter of the breadboard and lift gently.

It is vitally important to establish a general wiring technique that will aid in circuit debugging. Some recommended guidelines for the connection of wires to the breadboard will be discussed. By adhering to these guidelines, an ordered, neat circuit will be constructed which will simplify the troubleshooting of the circuit.

The first and most important precept to adhere to is neatness. First, it is a good practice to route wires around, not over, ICs. Second, cover as few unused holes as possible with wiring. Third, color code your wiring; for example, red for power, black for ground, green for data paths, and yellow for control paths. Fourth, keep the wires as short as possible. Two problems can result from not using short wires. First, it becomes almost impossible to debug a nonfunctioning circuit because of the "jungle" construction. Second, the IC's signals may radiate into other wires (crosstalk) causing faulty behavior. Therefore, it is recommended that a reasonable number of short, prestripped wires be prepared and retained for repeated use. If a wire end becomes bent, trim and restrip it.

In industry, a wiring technique called "wire-wrap" is commonly used in breadboarding. You may prefer to use wire-wrap components since they provide a strong and reliable connection and result in a neat circuit layout.

The following sequencing scheme will also aid in an ordered construction of a circuit. First, wire in the power and ground leads using the previously described color-coding technique. Second, wire regular buses that contain common signals, such as a clock pulse. Third, wire the control signals. The advantage to this wiring method is that power and bus wires are the least likely to require change during debugging.

A well-planned chip placement plan is also important to avoid creating the "jungle" of wires on the breadboard and to ensure adequate space for all the chips to be put on the breadboard. Do *not* reverse the orientation of ICs to minimize wire lengths since it can cause confusion in wiring and debugging.

Prepare a detailed wiring schematic for each experiment. As wires are inserted into the circuit, mark them off on the detailed drawing. This is a systematic way to ensure a complete and correct wiring for a complicated circuit.

3.1 Design and Debugging Strategies

As the complexity of IC circuit increases, the probability of the circuit working on the first try decreases rapidly. The term "debugging" has been referred to a process of identifying and correcting faults in both hardware and software designs. The term "debugging" implies the circuit does not perform according to its specification. A specification of the circuit design requirement is, therefore, critical to the correctness of the circuit construction. If no specification exists on the circuit, any function it performs must be assumed correct.

A logical approach to a large circuit design is the following:

- understand the requirements provided for the design,
- write a design specification ensuring the specification fulfills all the needs of the requirements,
- design the circuit,
- implement the design on a testing environment to verify the design, and
- debug any problems detected. (Design and/or implementation errors should be corrected and appropriate documentation updated.)

Although the above steps should be followed, the documentation can be informal. Without any documentation describing the circuit functions, inputs, and outputs, and without notes for implementing the design, testing cannot proceed in a clear, logical manner. Many large organizations have discovered that time spent on developing clear, complete requirements and design helps in testing, reduces debugging time, and reduces "design maintenance" required of circuits. If the design is implemented by a group as recommended for the design experiments of this book, the team members can inspect the designs produced to catch the possible errors before the circuit is implemented. Peer member inspection is one of the most effective ways to look for errors and to ensure product quality. As a good practice, you should complete a rough design drawing including part numbers and pin numbers for the circuit before building the circuit for any experiment in the book.

The types of documentation required for debugging are:

- a description of the circuit operation and notes on implementing the design developed from the output of the design step, and
- a procedure describing the logical steps to perform during the testing, which should be created before circuit testing and after circuit design.

It should be noted the above steps require that the circuit be designed first and implemented second. Although it is useful to build prototype circuits to study component action, you should never "design" a circuit using the implementation first and design last approach.

If you cannot "document" your circuit as to its function, inputs, and outputs, the circuit cannot be debugged successfully.

Hardware problems that an individual who has never implemented a circuit can expect to experience are:

- incorrect implementation of the circuit,
- incorrect wire and component connections,
- faulty components, and
- design problems.

You should double check the wiring against the circuit diagram and make sure that the right IC is correctly inserted before applying any power. Always check to be sure power and ground have *not* been wired together. Initially, the first-time implementor and experienced designer in a rush can expect he/she implemented the circuit incorrectly. Thus, if a circuit is not performing correctly, check the implementation. Additionally, the connections between components and wire may not be a true physical connection. Where component sockets and breadboard devices are used, the connections should be checked with a measuring device. After you have gained experience in implementing circuits, broken wires and poor connection problems become less frequent.

If the implementation and connections are correct, then either the design or the components are faulty. A faulty component can exhibit a constant output regardless of changes on its input. To debug the circuit, you must have a method for detecting logic levels and signals. Complex circuits require logic analyzers, oscilloscopes, and possibly microprocessor development systems. Experiments requiring implementation in this book need only the simple LED lamp or commercial logic probe to debug. Make sure that the testing tools (such as input switches, LEDs, and logic probes) themselves are in proper working condition. The use of debugging instrumentation is usually included with the instrument and is not covered in the book.

Multiple-Stage Circuit

A circuit can have many stages. A stage in a circuit is generally considered gate(s) in SSI chips (or digital functions in MSI and LSI chips) that are logically at the same level. A logic circuit stage is considered operating correctly only if its output is correct for all the inputs to the circuit. There are three different approaches to debugging multiple-stage circuits.

- Start at the first stage and progress stage by stage forward until a fault is detected. Once the fault is resolved, repeat the process through the final stage.
- Start at approximately the middle stage. If the middle stage is functioning correctly, work forward to the end of the circuit until a fault is found. Then fix errors and work forward until all stages are functioning correctly. On the other hand, if the middle stage is functioning incorrectly, work backward until a stage is found functioning correctly. Then fix errors and work forward until all stages are functioning correctly.
- Start at the final stage and progress stage by stage backward until a stage is found functioning correctly. Then progress forward resolving faults in stages containing errors.

The method used is generally determined by personal preference. However, the following information may help you in determining what method to use. If the circuit is a new design and you are not confident in the design and implementation, use the first method. If you are fairly confident in the design of the circuit and used the wiring techniques mentioned earlier, use the second method. An iteration of the second method is to progress in the desired direction in multiples of half steps until the problem is found. If you are very confident in the design and wiring, use the third method: start at the last stage and work back until a stage is functioning correctly and fix the stage in error.

3.2 Debugging Combinational Circuits

The initial experiments in this book are combinational circuits. Combinational circuits do not contain storage elements; the combinational circuit's output, therefore, reflects its current inputs only. If the output is in error, check the input conditions first. If the inputs are correct, check circuit implementation. Next, proceed per suggestions under "Multiple-Stage Circuit" section of this chapter. If additional information is required, J. D. Greenfield's book (in Reference 4 of Logic Design in the Bibliography) has a good discussion about the debugging approaches. The following is an example developed to illustrate a debugging approach.

Example 1

In Experiment 1 of Chapter 4, you are asked to implement the following majority circuit.

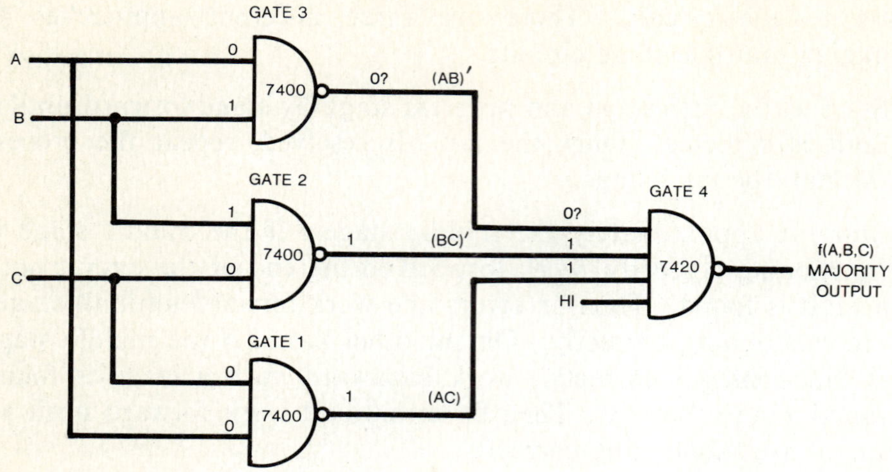

Let us assume that the circuit has the following bug after it is constructed: it produces a high output in response to an input (A = 0, B = 1, or C = 0). Locate the problem.

Solution:

First, use the schematic diagram of the majority circuit to check that each point is correctly connected and properly terminated in the socket. Remember that every connection has two sides and both sides should be checked to catch faulty wiring problems. If no problem is found, set ABC equal to 010 and check the logic levels throughout the circuit. The results are also shown in the diagram. An examination of the circuit reveals that gate 3 is not operating as it should. The 0 input to the top leg of the NAND gate 3 should produce a 1 output. A question-marked value is used to indicate an unexpected result in the diagram.

The problem definition indicates the component is not functioning correctly. At this point, the wiring has been verified, and component orientation and connections should have been verified. If not, verify component orientation and component/wire connections at this time. Before condemning the component, ensure that its pins are not bent. (You should exercise care when inserting and removing ICs and you should use pliers to straighten bent pins.) Reinsert if bent pins are suspected and retest. If the above is all correct, the component is at fault. The bad component should be located and replaced. The faulty IC should be properly marked or thrown away to prevent causing the same type of errors again.

3.3 Debugging Sequential Circuits

Most of the experiments in this book contain sequential circuits. Sequential circuits contain storage devices such as memories, flip-flops, registers, etc., in addition to combinational circuits. Since time is a factor in sequential circuits, you should generate a timing diagram during the design phase (or before testing if you are performing only the implementation phase such as the instructional experiments).

For complicated sequential circuits, some errors are not easily reproducible due to the timing problem. It would then be very difficult to locate the errors. If you try and still cannot locate a bug in your wired circuit, consult with your laboratory teaching assistant or instructor if the experiment is performed in a supervised laboratory. It is important for you to have a clear design document and neatly wired circuits if you really expect anyone to help you debug successfully.

The large software and hardware projects today must follow certain design methodologies to develop high-quality products. One thing to keep in mind is this: a good software or hardware design cannot be achieved by testing and debugging a design that is inherently bad; it can be achieved only by the process of engineering and building it. As the complexity of the circuit increases, Computer Aided Design (CAD) and automatic testing and verification techniques are often used to locate the design problems before the implementation. For a large project, more than 40 percent of the effort is usually expected to be spent in testing and debugging. One must go through this painstaking debugging process before a "working" implementation can be realized.

4

Instructional Experiments

Instructional experiments demonstrate the basic principles and electrical characteristics associated with digital circuits. (See References 2, 3, 7, 8, 10, 12, and 13 of Logic Design in the Bibliography.) Another purpose of these experiments is to supplement information not necessarily found in a standard logic design textbook. You will also learn to debug circuits and solve implementation problems which may not be encountered with pencil and paper designs. If you are experienced in IC experiments, the instructional experiments may not be challenging to you since they provide the circuit diagrams for you. You are encouraged to concentrate on the more challenging design experiments, design exercises, and design projects.

All of the design, circuit diagram, and wiring schematics of instructional experiments have been thoroughly tested by the students of the computer architecture courses at Illinois Institute of Technology during the past 3 years. Many sophisticated design projects can be carried out once you know the basic building blocks used in the instructional experiments. There are six experiments in this chapter:

Exp. 1 — Majority Circuit and Parity Detector

Exp. 2 — Binary-to-Decimal Decoder

Exp. 3 — Flip-Flops

Exp. 4 — Parallel/Load and Shift Register

Exp. 5 — Binary Coded Decimal (BCD) Counter and 7-Segment Display

Exp. 6 — Random Access Memory

If you are not familiar with digital systems, you should implement all six experiments. If you are already familiar with combinational circuits, you may skip Experiments 1 and 2. Experiments 1 through 5 should take 1 week each to complete. Experiment 6 should take 1 to 2 weeks to complete.

Each experiment has the complete logic wiring schematics for implementing the circuit. The purpose of these diagrams is to help you understand the implementation of logic circuits by IC chips. These wiring diagrams will not be supplied for the design experiments in the book. In a logic wiring schematic, if leads (or wires) cross together and are intended to be connected, a dot will be marked at the junction.

4.1 Experiment 1 - Majority Circuit and Parity Detector

This is a very simple experiment that will allow you to get familiar with the trainers and other equipment. However, you should realize that preparation for the experiments in the book is required before starting the implementation on the breadboard.

There are three parts in this experiment. Part A introduces a 3-input majority circuit which was used as a debugging example in Chapter 2. Part B requires the implementation of a parity detector circuit. Part C is a sample laboratory report for the experiments in Parts A and B.

4.1.1 Part A - Three-Input Majority Circuit

Objectives:

To implement a 3-input majority circuit whose output is HIGH if and only if two or more inputs are HIGH. Otherwise, the output is LOW.

Materials:

1. One 7400 quad 2-input NAND gate.
2. One 7420 dual 4-input NAND gate.

Equipment:

A logic trainer.

Procedures:

1. Construct the 3-input majority circuit as shown in Figure 4-1.
2. The circuit wiring diagram is shown in Figure 4-4 for reference.
3. Demonstrate the operation of the circuit to the laboratory teaching assistant if it is performed in a supervised laboratory.

Questions:

1. Show two techniques for using a 4-input NAND gate as a 3-input NAND gate.

2. Find the TTL-type number for a 4-input AND gate. Can you find any 5- or 6-input AND gates in any manufacturer's data book?

3. Draw the logic diagram showing how a 5-input OR gate can be implemented using 2-input OR gates.

4. Under what circumstances might it be necessary to use several gates to take the place of a single multiple-input gate?

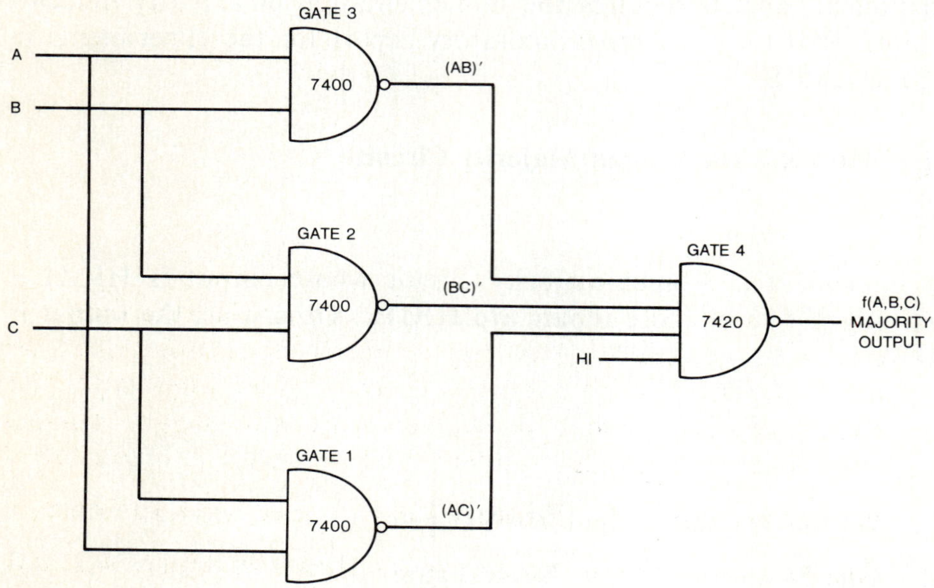

Figure 4-1. *A 3-Input Majority Circuit*

4.1.2 Part B - Parity Detector Circuit

Introduction:

Data communication between two parties requires the correct delivery of data characters between them. Since communication noise can corrupt the data being transferred, it is important to check the correctness of the data. To ensure that the data sent is the same as the data received, a parity bit can be added to the character bits at the sending side to be checked for transmission errors at the receiving side. This bit can cause the character to be even (total number of "one" bits being even) or odd. If an error is detected, the retransmission of the data may be required to recover from the error.

This experiment is concerned with a parity detection circuit that can be used as part of a parity generation circuit or parity error detector circuit. Experiment 10 also covers the error-detection protocol circuits that can do parity checking for a block of characters.

Objectives:

To implement a parity detector circuit using exclusive-OR gates.

Materials:

One 7486 quad 2-input exclusive-OR chip.

Equipment:

A logic trainer.

Procedures:

1. Prepare an implementation procedure for constructing the circuit using Figure 4-2 as the input. This document can be informal, but the document is necessary to implement the circuit. Figure 4-2 must be supplemented with the gate's input pin numbers, output pin numbers, and the power pin numbers.
2. Prepare a truth table. Figure 4-3 can be used as an example.
3. Set the input switches to the states shown in Figure 4-3 and record the results observed for each state.
4. The wiring schematic diagram is shown in Figure 4-5.

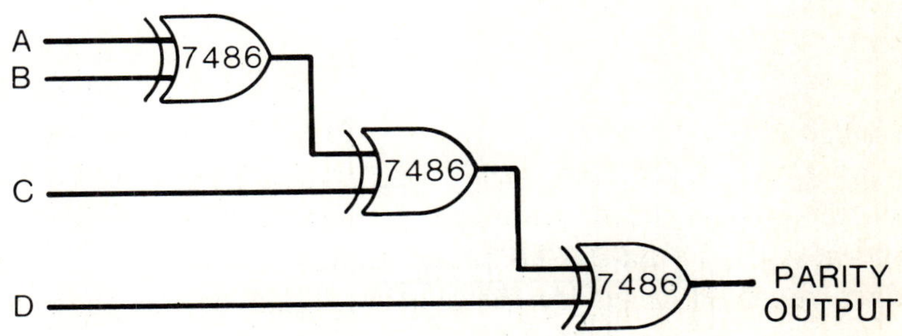

Figure 4-2. *Parity Detector Circuit*

Questions:

1. Show that an equality gate can be constructed by inverting the output of an exclusive-OR gate.
2. Explain how the circuit can generate parity and how the circuit can verify parity.

CODE GROUP	PARITY ODD-EVEN	NUMBER OF ONES IN CODE GROUP	PARITY OUTPUT
A B C D	PARITY	NO. OF 1's	
1 0 0 0			
0 1 1 0			
1 1 0 1			
0 0 0 1			
0 0 1 1			
1 0 0 1			

Figure 4-3. *Parity Output Table*

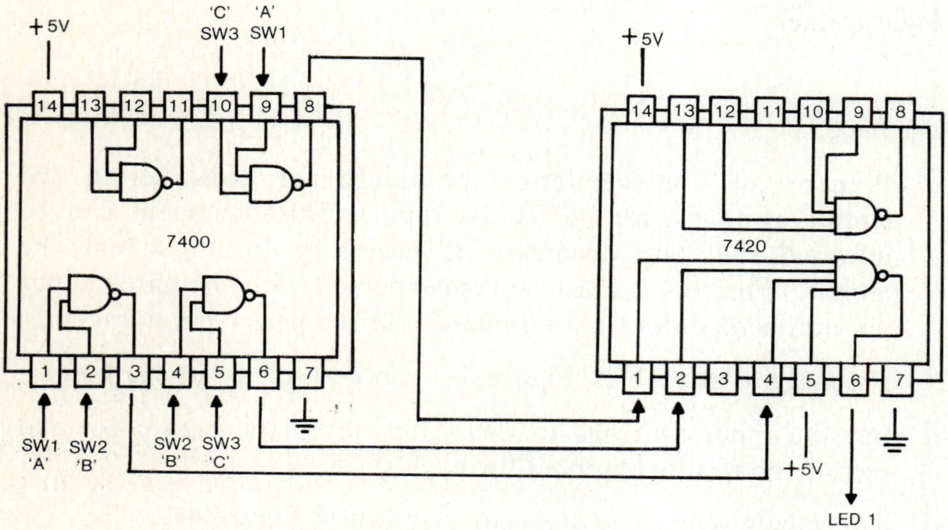

Figure 4-4. *Wiring Schematic of the Majority Circuit*

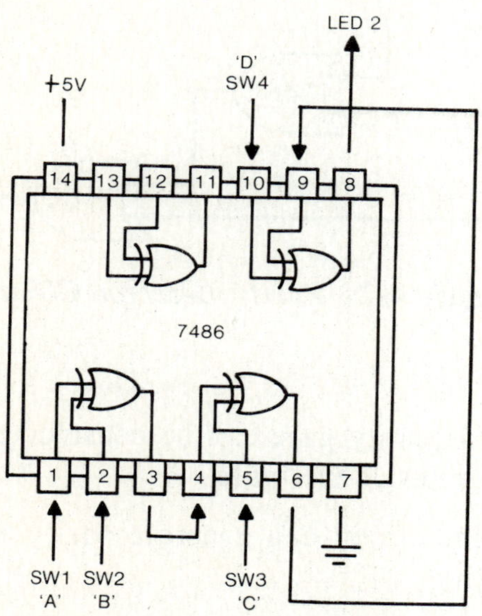

Figure 4-5. *Wiring Schematic of the Parity Detector Circuit*

4.1.3 Part C - A Sample Laboratory Report for Experiment 1

Write-Up Formats for Laboratory Report

If the book is used for a supervised laboratory course, a laboratory report should be written for each experiment upon its completion.

One objective of the laboratory report is to give you experience in writing good technical reports and in the preparation of experimental results. All laboratory reports should contain a clear presentation of the results obtained from the laboratory work. All reports should be neatly written or typewritten on one side of the page and bound together. The following will serve only as a sample format for writing a report for a laboratory class.

1. Use a title page, as shown below:

Class Name, Experiment Number	
Experiment Title	
Lab Time	Student's Name
Date Submitted	Lab Assistant

2. Provide a statement of the laboratory purpose and objectives.
3. Provide all necessary truth tables, state tables, and circuit diagrams for simple circuits, or block diagrams for the more complex circuits.
4. Include all observations that were made during the design or the laboratory phase of the experiment. Do not use a "1" or a "0" to indicate the states of the display LEDs. Instead, use the terms which directly describe these observations. For example, an LED is ON/OFF, Bright(B)/Dark(D), or HIGH/LOW.
5. Provide a comparison between the expected and observed results. A summary of unexpected events should be reported and discussed.
6. Give answers to all questions assigned by the instructor in the order specified.

A sample report for Experiment 1 is illustrated in the following as an example.

A. Three-Input Majority Circuit

Objectives:

To implement a 3-input majority circuit whose output is high if and only if two or more inputs are high and whose output is low otherwise.

Materials:

One 7400 quad 2-input NAND gate and 7420 4-input NAND gate.

Equipment:

A logic trainer.

Procedure and Observations:

The 3-input majority circuit truth table was first listed along with the associated function definition and Karnaugh map. A sum of products form of the equation is obtained by using the Karnaugh map. A simplified function, F, in terms of product of sums form is obtained by applying DeMorgan's Theorem. The derivation of the function for the majority circuit is shown in the following diagram (Figure a).

$F(A,B,C) = \Sigma\ (3,5,6,7)$

(a) Output Function

A	B	C	F(A,B,C)
0	0	0	0
0	0	1	0
0	1	0	0
0	1	1	1
1	0	0	0
1	0	1	1
1	1	0	1
1	1	1	1

(b) Truth Table

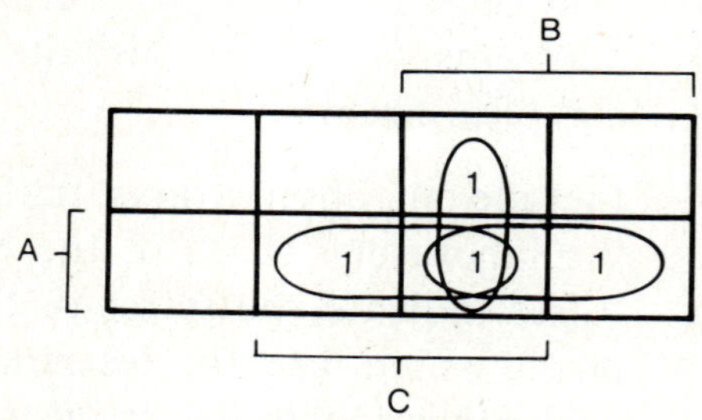

(c) Karnaugh Map

$F = AB + AC + BC$

(d) Minimum Sum-of-Product Function

Figure a. *Derivation of the 3-Input Majority Circuit Using Only NAND Gates*

A truth table comparison between the expected and observed results in the laboratory is performed and they were consistent for all eight test cases as shown in the following diagram (Figure b).

A	B	C	(AB)′	(BC)′	(AC)′	f(A,B,C)	OBSERVED RESULT
0	0	0	1	1	1	0	LOW
0	0	1	1	1	1	0	LOW
0	1	0	1	1	1	0	LOW
0	1	1	1	1	0	1	HIGH
1	0	0	1	1	1	0	LOW
1	0	1	1	0	1	1	HIGH
1	1	0	0	1	1	1	HIGH
1	1	1	0	0	0	1	HIGH

Figure b. *Truth Table for Majority Circuit*

Summary Questions:

1. Two techniques for using a 4-input NAND gate as a 3-input NAND gate are as follows:

    a. The first method would simply require the fourth unused input of the 7420 NAND gate to be constantly set HIGH. This HIGH setting would have no effect on the resulting NAND values from the other three inputs.

    b. The second method would require the result of any of the three outputs from the quad 2-input 7400 circuit to be doubly connected to two inputs of the dual 4-input 7420 circuit.

2. The list of type numbers associated with the dual 4-input positive AND gate according to the TTL data book is the following:

    SN 54H21(J) SN 54LS21(J,W) SN 74H21(J,N)
    SN 74LS21(J,W) SN 54H21(W)

    Type numbers for any 5- or 6-input AND gates could not be found in any manufacturer's data book.

3. A 5-input OR gate can be implemented using 2-input OR gates in the following manner:

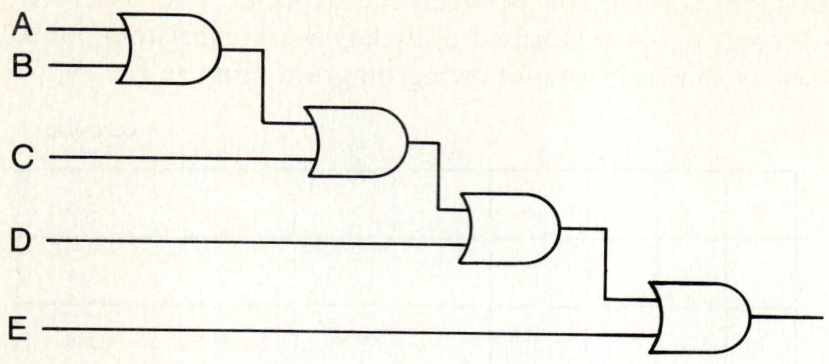

4. Circumstances which might require the use of several gates to take the place of a single multiple-input gate are as follows:

   a. The required multiple-input gate might not currently be manufactured. For example, 5-input AND gates are not manufactured. A gate may not be manufactured due to low demands or large numbers of wasted IC pins.

   b. When an intermediate result is needed, several gates may be taken to replace a multiple-input gate.

B. Parity Detector Circuit

Objectives:

To implement a parity detector circuit utilizing exclusive-OR gates.

Materials:

One 7486 quad 2-input exclusive-OR IC.

Equipment:

A logic trainer.

Procedure and Observations:

The circuit diagram and truth table for the exclusive-OR parity detector circuit was generated.

A truth table comparison between the expected and observed results shows that they were equivalent for all 16 test cases. The circuit and table are shown in the following diagram (Figure c). Table entries required are prefixed with an asterisk.

Summary Questions:

1. $(A\overline{B} + B\overline{A}) = (\overline{A} + B)(\overline{B} + A) = \overline{A}\,\overline{B} + AB$ by DeMorgan's Theorem.

2. The answer is not included in this sample report and is left as an exercise for you.

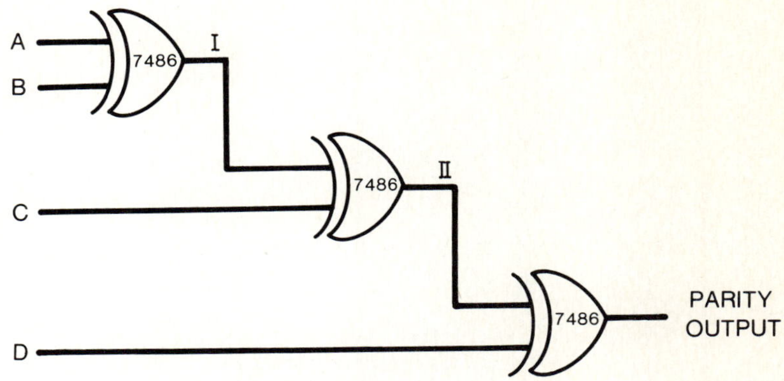

(a) Circuit Diagram

	A	B	C	D	I=A⊕B	II=I⊕C	II⊕D	PARITY	NO. OF 1's	OBSERVATION RESULT
	0	0	0	0	0	0	0	EVEN	0	LOW
★	0	0	0	1	0	0	1	ODD	1	HIGH
	0	0	1	0	0	1	1	ODD	1	HIGH
★	0	0	1	1	0	1	0	EVEN	2	LOW
	0	1	0	0	1	1	1	ODD	1	HIGH
	0	1	0	1	1	1	0	EVEN	2	LOW
★	0	1	1	0	1	0	0	EVEN	2	LOW
	0	1	1	1	1	0	1	ODD	3	HIGH
★	1	0	0	0	1	1	1	ODD	1	HIGH
★	1	0	0	1	1	1	0	EVEN	2	LOW
	1	0	1	0	1	0	0	EVEN	2	LOW
	1	0	1	1	1	0	1	ODD	3	HIGH
	1	1	0	0	0	0	0	EVEN	2	LOW
★	1	1	0	1	0	0	1	ODD	3	HIGH
	1	1	1	0	0	1	1	ODD	3	HIGH
	1	1	1	1	0	1	0	EVEN	4	LOW

★ *INDICATES CODE GROUPS SHOWN IN FIGURE 4b OF THE LAB DESCRIPTION*

(b) Observed Output Table

Figure c. *Circuit Diagram and Truth Table for Parity Detector Circuit*

4.2 Experiment 2 - Binary-To-Decimal Decoder

Introduction:

When a computer has completed an operation, the answer is usually coded in binary form in the memory storage. For example, a 3-bit operation code of a computer instruction can be used to represent eight different operations. Since it is easier to understand a display of combinations with one bit position representing each combination, it is helpful to learn a method of decoding coded information. One method is to use a NAND gate decoder as shown in Figure 4-6. The truth table, shown in Figure 4-7, indicates the operation rules of the decoder. Note that the bar symbol on the top of a variable is another way of designating negation.

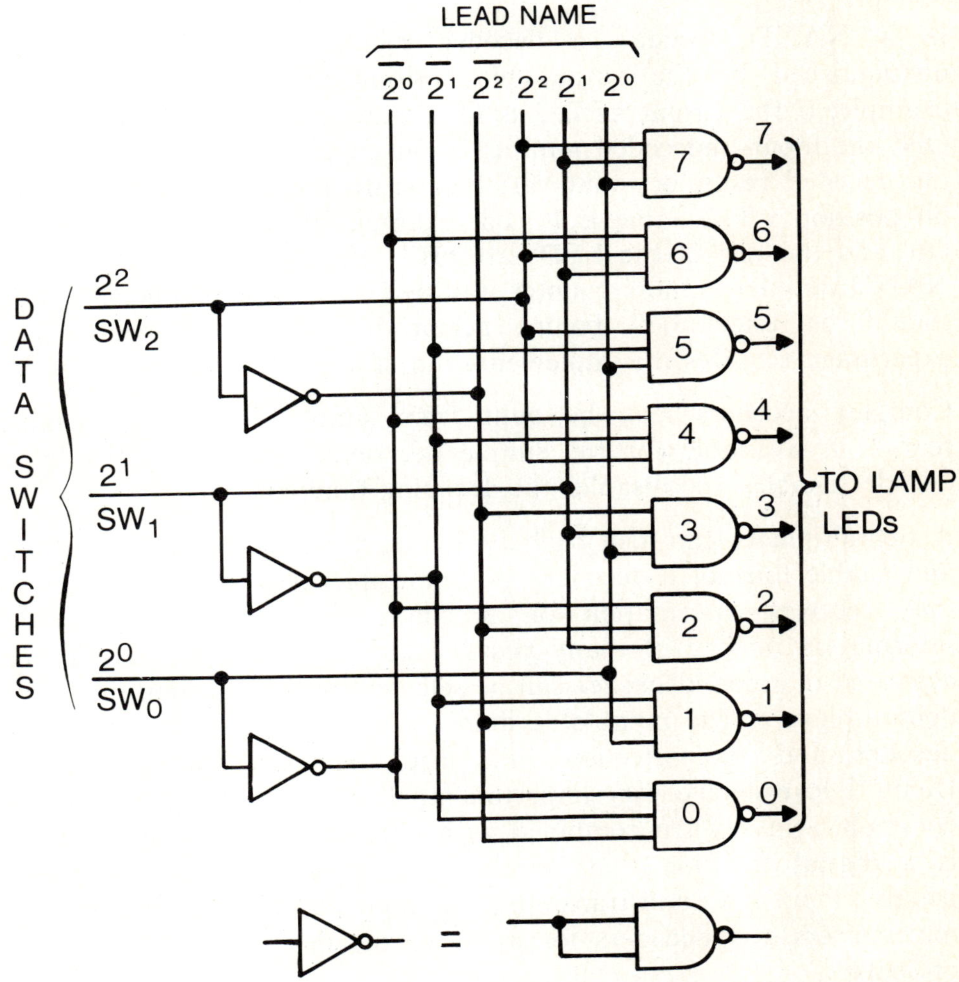

(The Inverter Implementation from a NAND Gate)

Figure 4-6. *Binary-to-Decimal Decoder Circuit*

2^2	2^1	2^0	$\overline{2^2}$	$\overline{2^1}$	$\overline{2^0}$	DECIMAL
0	0	0	1	1	1	0
0	0	1	1	1	0	1
0	1	0	1	0	1	2
0	1	1	1	0	0	3
1	0	0	0	1	1	4
1	0	1	0	1	0	5
1	1	0	0	0	1	6
1	1	1	0	0	0	7

Figure 4-7. *Binary-to-Decimal Decoder Operating Truth Table*

In a NAND decoder, a decoded result of input variables is distinguished by the output bit position whose value is 0. For example, if the output LED 2 is OFF and the rest of the LEDs are ON, the input undecoded number is binary 2. In an AND decoder, the decoded result of input variables is distinguished by the output bit position whose value is 1. For example, if the output LED 0 is ON and the rest of the LEDs are OFF, the input is binary 0. The NAND decoder is more counter intuitive than the AND decoder and should be understood before performing the experiment. The experiment requires the implementation of a NAND decoder only.

Complex decoders may be built from simple ones if the enable feature is available for the simple decoders. The enable line of a decoder enables and disables the decoding function.

A demultiplexer can also be built by rearranging the input, output, and enable lines of a decoder. A demultiplexer is a digital function that receives information on a single line and transmits the information on one of many possible output lines. The output line activated in a demultiplexer can be selected by the selection lines of a demultiplexer. The input of a demultiplexer is the enable line of a decoder and the selection lines are the input of a decoder. Demultiplexers have very important applications in the field of data communications when combined with multiplexers. The multiplexer is a digital function that selects information from one of many possible input lines and transmits to a single output line. We will use multiplexers and decoders for many of the design experiments and projects.

Objectives:

To study the behavior of a binary-to-decimal decoder circuit.

Materials:

1. One 7400 quad 2-input NAND gate.
2. Four 7420 dual 4-input NAND gates.
3. Four 150-ohm resistors.
4. One 7404 inverter (optional).
5. Eight LEDs are highly desirable for the observation of this experiment.

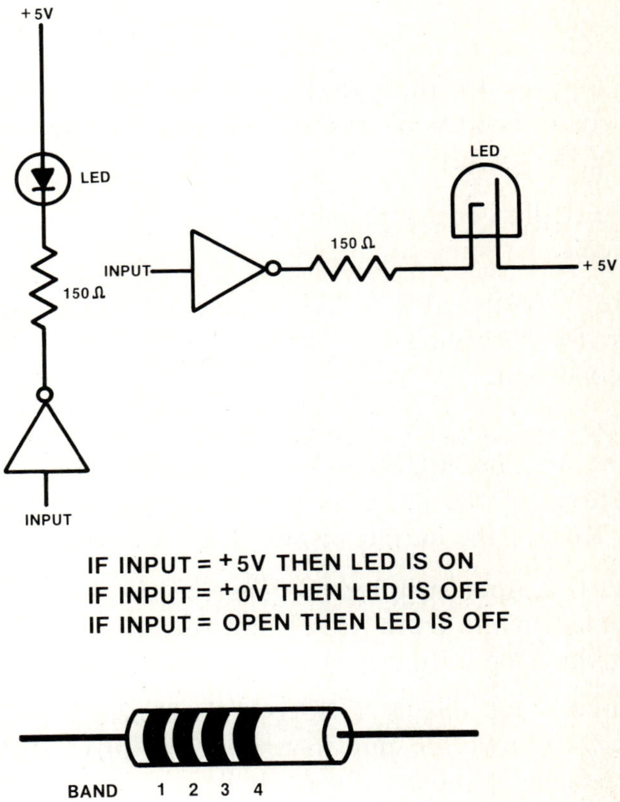

150Ω BAND 1 = BROWN, BAND 2 = GREEN, BAND 3 = BROWN

10KΩ BAND 1 = BROWN, BAND 2 = BLACK, BAND 3 = ORANGE

Figure 4-8. *LEDs and Resistors*

One method of building extra indicators using LEDs on the trainer is shown in Figure 4-8. LEDs are small diodes that are illuminated when current is passed through them. When an LED is ON, it develops a forward voltage of 1.6 volts, regardless of its current. Typical LEDs are specified to operate at 20 mA, but they will glow over a wide current range. For use with TTL circuits, a 150-ohm resistor is placed in series with the diode to limit the current and drop the excess voltage. Figure 4-8 illustrates a general connection technique of an LED circuit. It also shows the technique for identifying the 150-ohm and 10K-ohm resistors that are used in the book.

Equipment:

A logic trainer.

Procedures:

1. Study Figures 4-6 through 4-10 to develop an understanding of the circuit operation and to obtain information on implementing the circuit.
2. Prepare a plan for implementing the circuit and then connect the circuit.
3. Prepare a truth table with all possible inputs and provide a column for recording the output display LED status for each input condition.
4. Initialize the circuit input variables by setting the input switches to the LOW (OFF) position. Be sure that the orientations of the switches and the input binary bits are the same. Record the output display LED status in your table.
5. Change the input switches to generate the next input condition (as defined in the truth table) and record the new LED status. Repeat this step until all input conditions have been generated.
6. Remember that this is a NAND decoder. You may want to use the 7404 inverter chip to convert the NAND decoder to an AND decoder if the chip is available.

You may want to read a TTL data book and learn how the 74154 4x16 decoder chip works. The circuit diagram of the 4x16 decoder is shown in Figure 4-9.

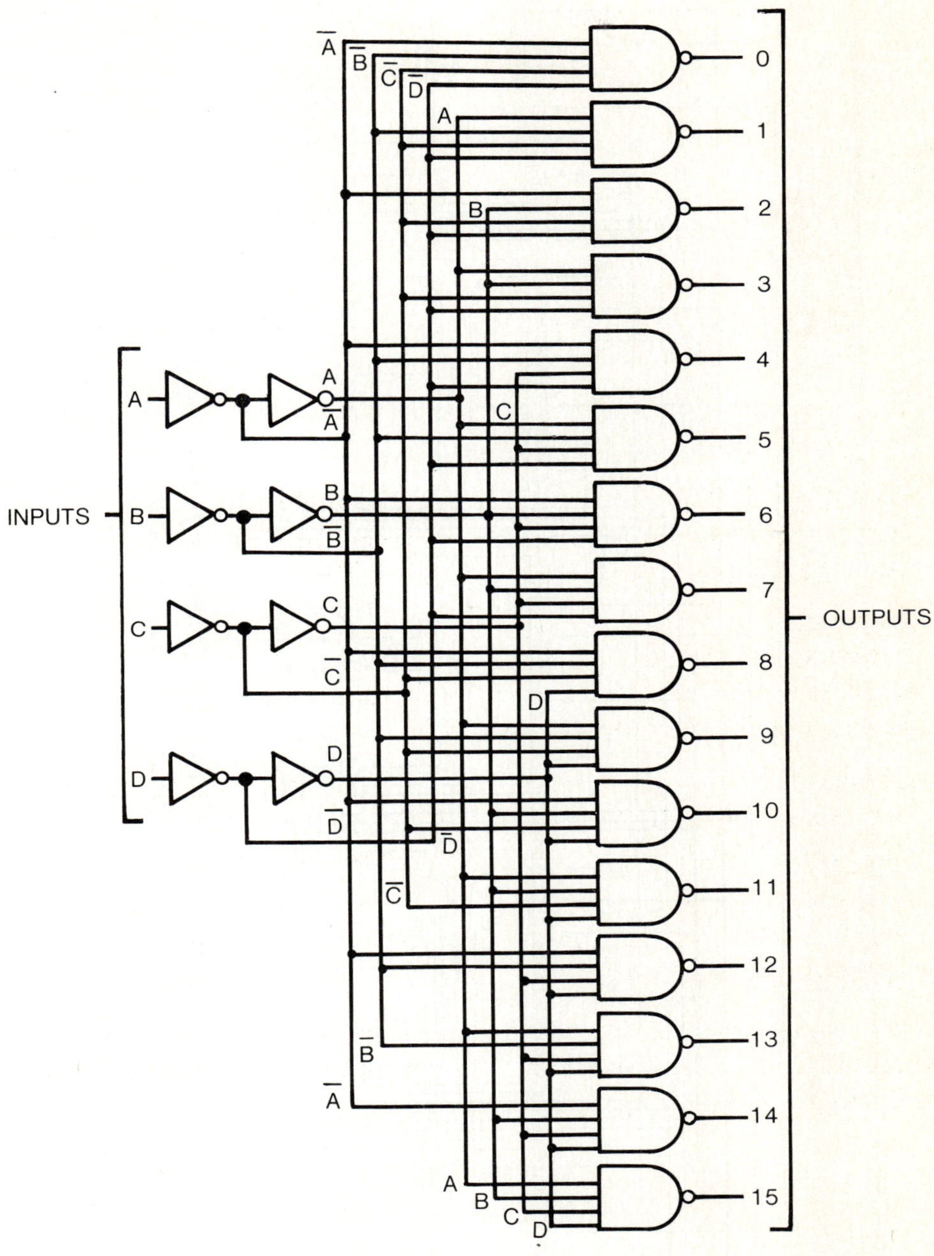

Figure 4-9. *4 X 16 Decoder Circuit*

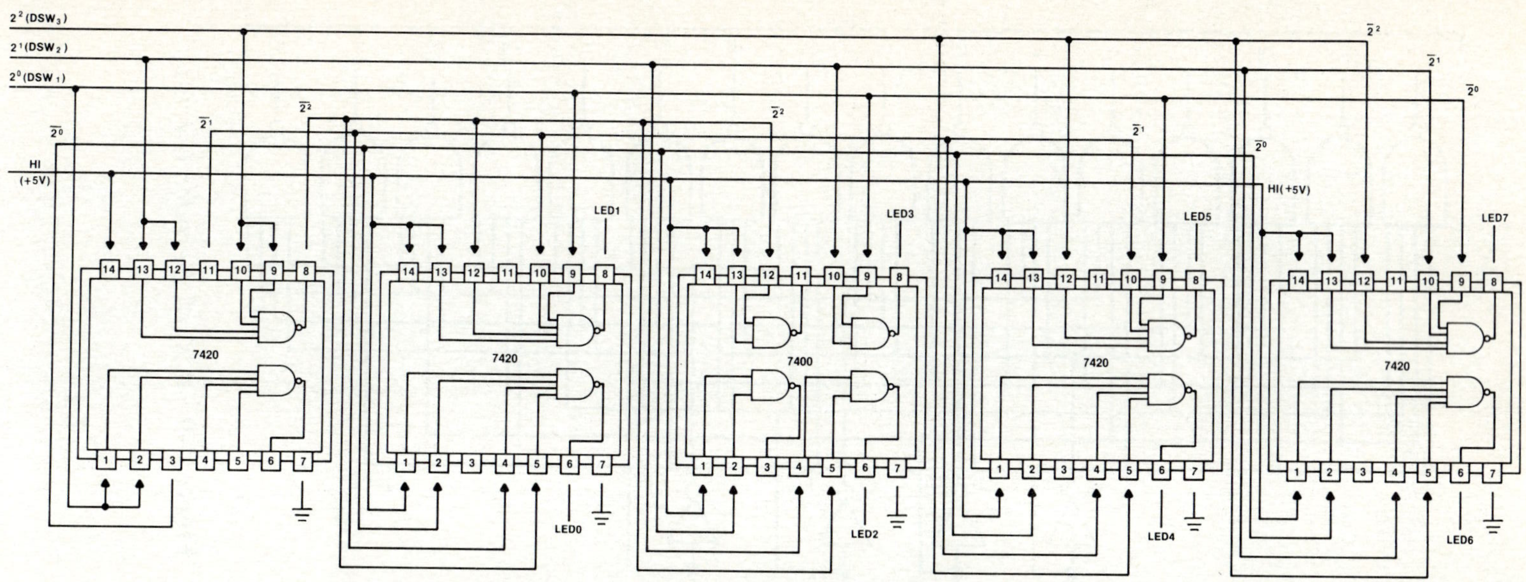

Figure 4-10. *Circuit Kit Schematic for the Binary-Decimal Decoder*

4.3 Experiment 3 - Flip-Flops

Introduction:

One of the most basic components used in sequential circuits is the Flip-Flop (FF). There are several types of FFs that can be built on a simple latch circuit. An FF is a component that can hold one bit state even in the absence of inputs. Its state change depends not only on the current inputs but may also depend on its current state. With the FF components, many types of sequential circuits can be created. One of the most commonly used sequential circuits is the register, which is a combination of several FFs. Registers are more expensive memories than other types of memories. They are mainly used in logic designs where speed is a more important factor than cost.

An FF may have an asynchronous (unclocked) or synchronous (clocked) control. A clocked circuit means that the input signals are accepted by the circuit only at the clocked instant of time. An unclocked circuit means that the input signals are accepted at any instant of time when a control signal is presented. In this book, we do not specifically distinguish the clocked and unclocked circuits. Most of the circuits are clocked. However, during the experimentation, the clock is usually generated by your activation of the control pulse from the logical switch.

This experiment introduces the FF circuits in five parts:

A. the latch circuit,

B. the Reset Set (RS) FF,

C. the Data (D) FF,

D. the Toggle (T) FF, and

E. the master-slave JK FF.

Objectives:

1. to become familiar with FF concepts,
2. to implement and observe the behavior of various FFs, and
3. to understand the concept of leading-edge and trailing-edge triggering, and clocked and unclocked control.

Materials:

1. One 7420 dual 4-input NAND gate.
2. Two 7400 quad 2-input NAND gates.

Equipment:

A logic trainer.

4.3.1 Part A - RS NAND Latch Circuit

Procedures:

Build a basic RS latch circuit as shown in Figure 4-11. This is the most primitive circuit to store information. Notice the notations used in the timing diagram. The SET and RESET are activated by negative pulses. The arrows indicate the cause and effect with the dot(s) representing the cause and the arrowhead representing the effect. Connect the inputs to the data switches and Q and Q' to the display LEDs. Obtain the state table experimentally. The circuit wiring diagram is shown in Figure 4-16 for reference.

The RS latch operates as follows:

1. In the timing diagram in Figure 4-11, we assume the latch was initially at RESET state (with Q at LOW and Q' at HIGH) when the power is first turned on. If the latch is not at RESET state, do not worry. You should still continue performing the rest of the procedures.
2. If SET input to pin 1 is LOW, it will cause Q HIGH. Since both the RESET input and pin 2 are HIGH, Q' and pin 1 are LOW.
3. If the SET input now goes HIGH, the latch **remains** in its SET state because the level at pin 1 remains LOW and both inputs to the lower NAND gate remain HIGH. This is the "memory" capability of an FF and a latch.
4. This condition continues until a RESET input goes LOW, which forces Q' HIGH. Now both SET and pin 1 are HIGH, causing Q to go LOW.
5. The latch remains in this reset condition until the next SET pulse goes LOW again.

Notice that if you set both of the SET and RESET inputs to LOW simultaneously, you can force both the Q and Q' outputs HIGH. After the SET and RESET both change to HIGH, the latch is at an indeterminate state since the new value of Q is not predictable. This is the reason that we prohibit the use of SET and RESET to HIGH simultaneously for an RS latch.

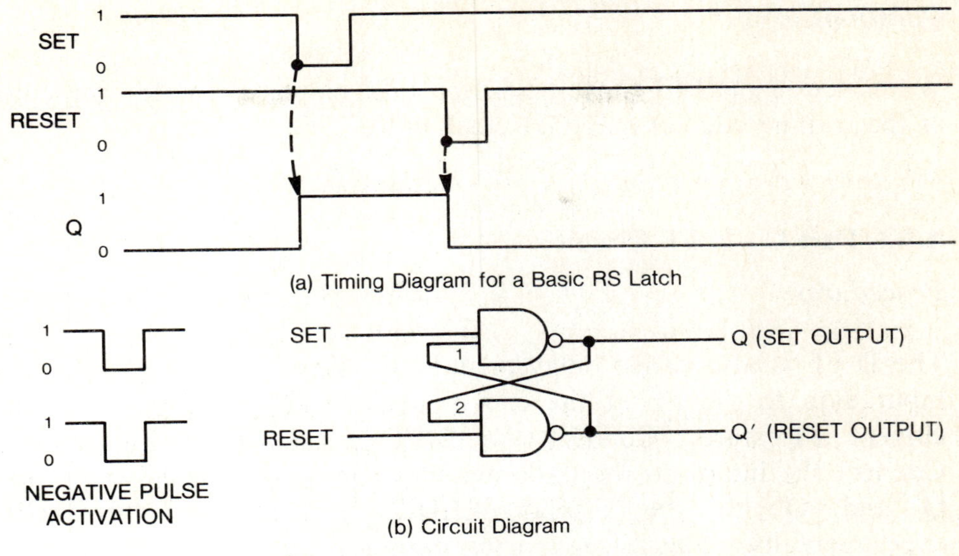

Figure 4-11. *A Basic RS Latch*

4.3.2 Part B - Basic NAND RS FF Circuit

Procedures:

Modify the circuit of Part A to form a clocked RS FF as shown in Figure 4-12. A clock control is now available and the circuit operates under positive pulses. Obtain the state table experimentally. Connect the inputs to the data switches, clock pulse to a logic switch, and Q and Q' to the display LEDs. The circuit wiring diagram is shown in Figure 4-16 for reference.

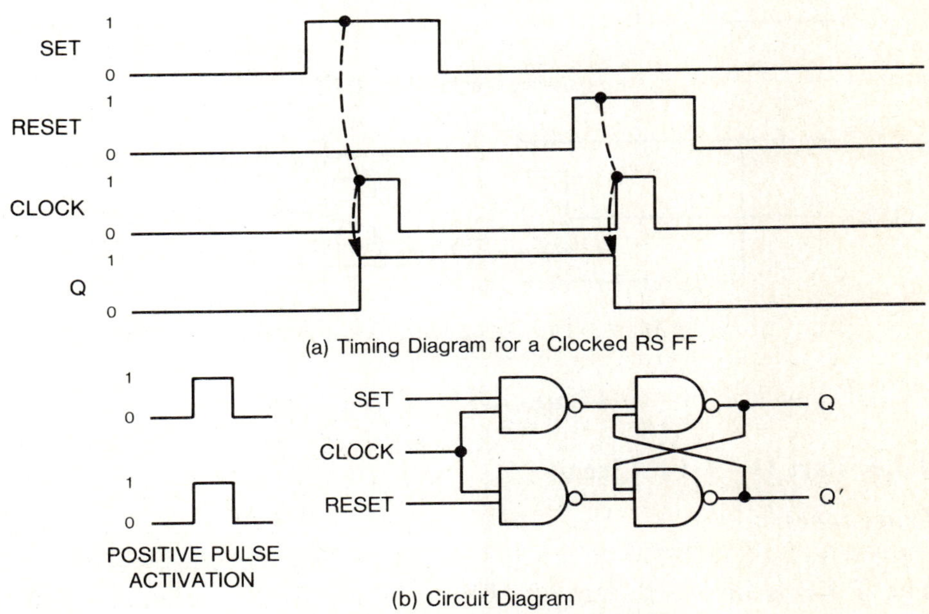

Figure 4-12. *A Clocked RS FF*

Questions:

When does an RS FF state transition take place, on the leading edge or the trailing edge of a HIGH clock pulse? Why?

4.3.3 Part C - D FF Circuit

Procedures:

The D FF is also called delay or data FF since it relays the current input data to the FF at the end of a clock pulse regardless of the current FF state. Connect the D FF as shown in Figure 4-13. Connect the inputs to the data switches, clock to a logic switch, and Q and Q' to the display LEDs. Obtain the state table experimentally. The circuit wiring diagram is shown in Figure 4-16 for reference.

Questions:

Are there any indeterminate states in the state table for this device?

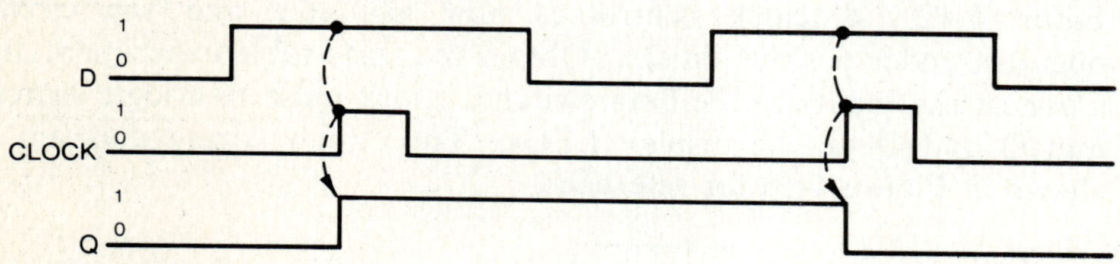

(a) Timing Diagram for a D FF

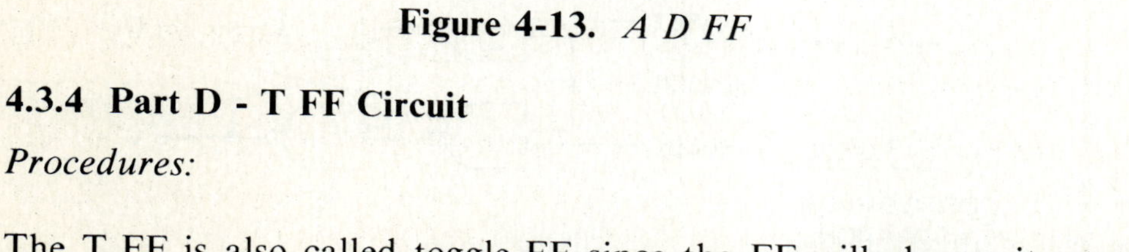

(b) Circuit Diagram

Figure 4-13. *A D FF*

4.3.4 Part D - T FF Circuit

Procedures:

The T FF is also called toggle FF since the FF will change its state when the input is high. Prepare your own wiring diagram using Figure 4-14 as the building circuit. Connect the components of the

T FF. Connect the input to a data switch, the clock to a logic switch, and the Q and Q' to the display LEDs. As in the D FF, develop the state table experimentally.

Questions:

Describe the different effect of the inputs on the D FF verses the T FF circuit from your observations.

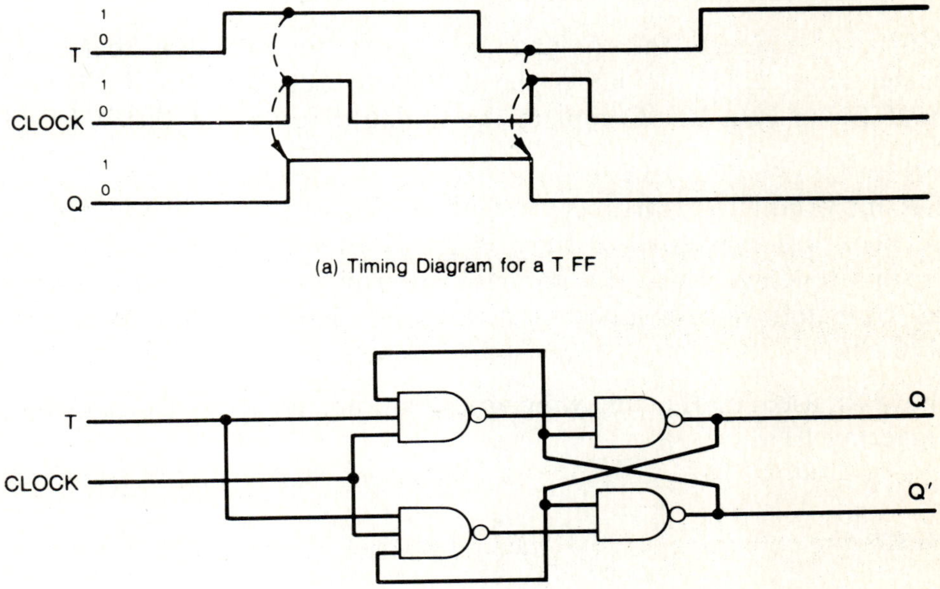

(a) Timing Diagram for a T FF

(b) Circuit Diagram

Figure 4-14. *A T FF*

4.3.5 Part E - Master-Slave JK FF Circuit

Triggering or activation of FFs can happen at the leading edge or trailing edge of clock pulses. The circuit in Parts A, B, C, and D are leading-edge triggering and the FFs change state when the triggering signal starts. The circuit for master-slave FF is trailing-edge triggering and the FF changes states when the triggering signal disappears. Activation of FFs at the trailing edge simplifies the interconnection of FFs in series and allows the same clock pulse to be used by all the FFs in a sophisticated digital system.

The circuit in Figure 4-15 realizes the goal of state transition on the trailing edge of a clock signal. Information is transferred during the leading edge of a clock signal and held there until the trailing edge of a clock signal. The slave responds at the trailing edge, allowing the new information from the master to become available at the slave's outputs.

Procedures:

Connect the master-slave JK FF as shown in Figure 4-15. Use a logic switch for the clock input; monitor points Q_m and Q_s via the display LEDs. Prepare a state table of outputs showing Q_m and Q_s for various combinations of J and K as the clock pulse moves from LOW to HIGH and back to LOW. Set J and K to HIGH. You can let the circuit operate synchronously and automatically by connecting the clock input to the 1-Hz clock pulse on the trainer. The circuit wiring diagram is shown in Figure 4-16 for reference.

You can connect the J and K inputs together to realize a master-slave T FF. What is the difference between this T FF and the T FF in Part D? Complete the timing diagram for this T FF in Figure 4-15.

It is important to remember that all of the FFs used in the rest of the book are master-slave FFs. The JK FFs in the 7476 chip are also master-slave FFs. An optional experiment for the student who has time left in the laboratory is to get familiar with using the 7476 chip.

The FFs used in the following questions are also the trailing-edge triggering FFs.

Questions:

1. Explain why all the input possibilities on the JK FF state table are valid.
2. What is the output function applied if 80 clock pulses are applied to the T FF circuit?
3. Realize a D FF given one T FF and any kind of gates.
4. Realize a JK FF given one D FF and any kind of gates.
5. Explain the function of clear and preset of the JK FF by examining the function table of the 7476 chip.
6. Two T FFs are connected together. The output of the first T FF is connected to the clock of the next, and the input of both T FFs are HIGH. The first T FF is connected to a 1-Hz clock. Draw a timing diagram for the output of both FFs.

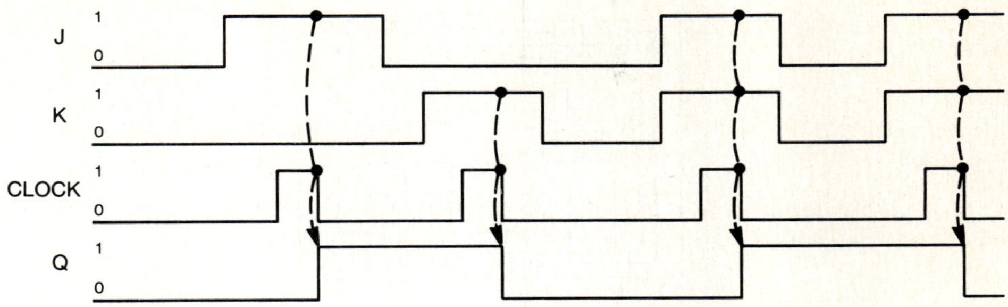

(a) Timing Diagram for a Master-Slave JK FF

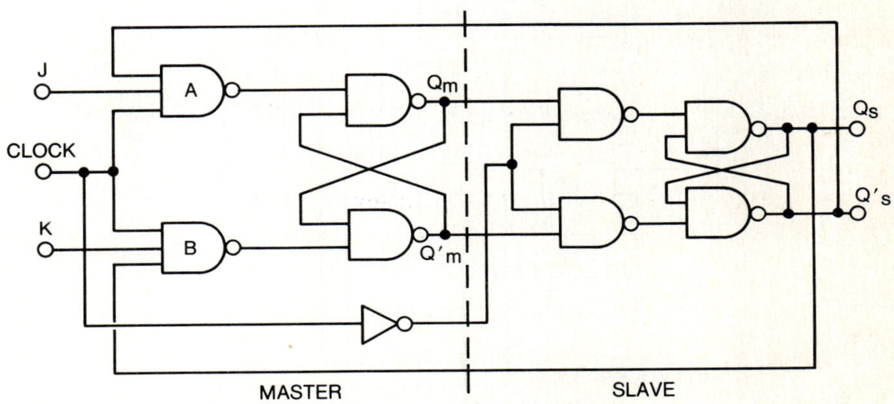

(b) Circuit Diagram of a Master-Slave JK FF

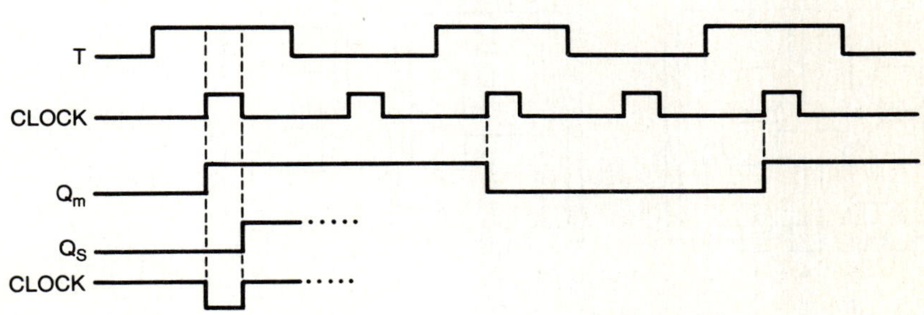

(c) A Partial Timing Diagram of a T Master-Slave FF

Figure 4-15. *A Master-Slave JK FF*

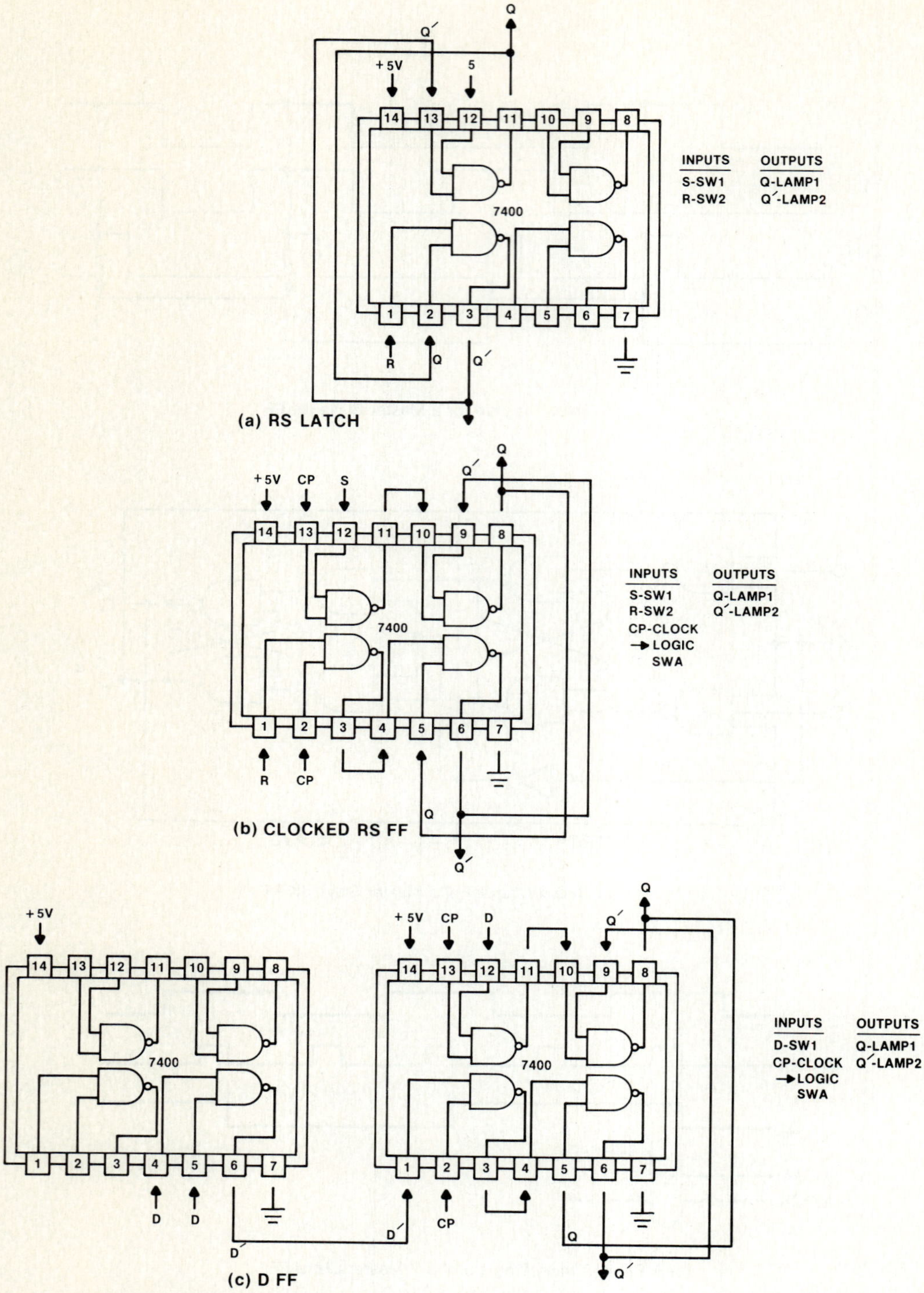

Figure 4-16. *Circuit Schematic for the FF Experiments*

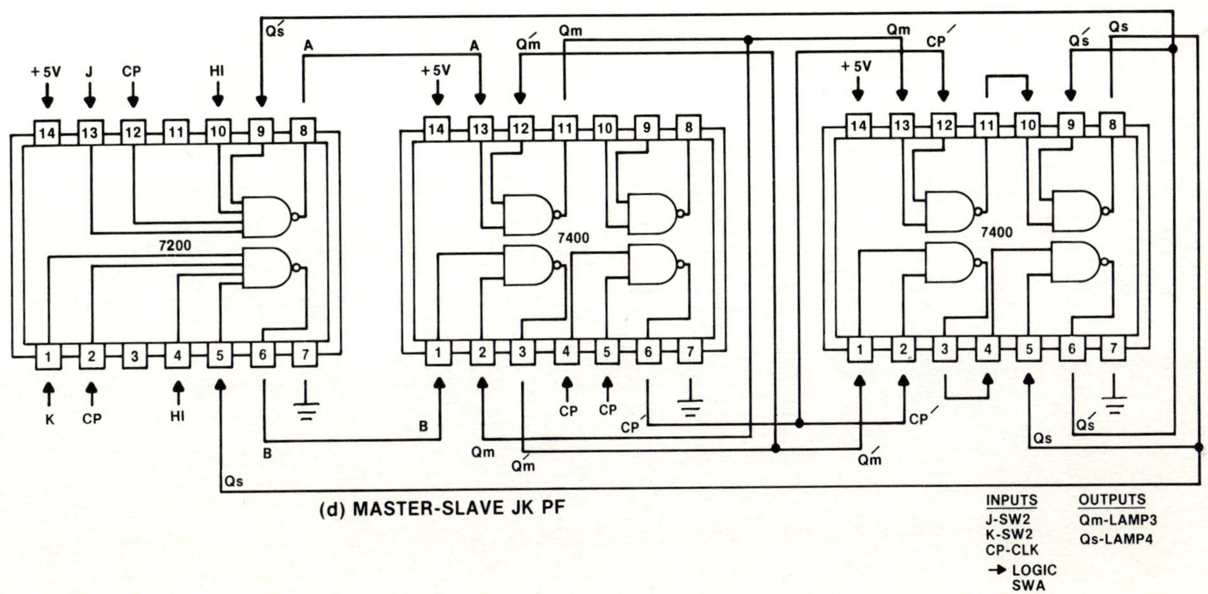

(d) MASTER-SLAVE JK PF

Figure 4-16. *(Cont.)*

4.4 Experiment 4 - Parallel/Load and Shift Register

Introduction:

Logic circuit components used in computers, input/output peripheral devices, and communication devices have either parallel or serial inputs and outputs. Interface circuits, such as the computer network interface design in Project 3 of the book, typically must accept one form of input (serial or parallel) and convert the data to the other form as an output. Some circuits such as adders may also be required to perform the serial shifting during the addition of each bit and be required to provide parallel inputs and outputs.

This experiment implements the different types of circuits from a parallel/load and shift register circuit. We organize the experiment in five parts:

A. right-shift register serial loading,

B. parallel loading,

C. circular-shift register,

D. Johnson counter, and

E. left- and right-shift register.

This experiment requires extensive wiring and is intended to serve as good practice for constructing large circuitry. The instructor of a supervised laboratory may choose to let you implement the experiment by using a single 7495 shift register chip described in the parts kit. You should consult the function table of the 7495 chip in Figure D-13.

Objectives:

1. to study the behavior of serial-in/serial-out circuits,
2. to study the behavior of serial-in/parallel-out circuits and parallel-in/serial-out circuits, and
3. to study the behavior of different types of shift registers.

Materials:

1. Two 7476 dual JK FF chips.
2. Five 7400 quad 2-input NAND gates.
3. Six data switches and one pulser switch are needed. One method to implement the additional data switches is by using a 10K-ohm resistor as shown in Figure 4-17. One end of the

resistor is connected to power (+5 volts). The other end of the resistor is connected to the desired lead such as Right Serial Input (RS) or Parallel Load Enable (PE) as shown in Figures 4-17 and 4-18. When a logic 0 is needed, the input side of the resistor is grounded. Similarly, when a logic 1 is needed, the input end of the gate is opened. The other method to implement the additional data switches is to use the optional DIP switch of the parts kit.

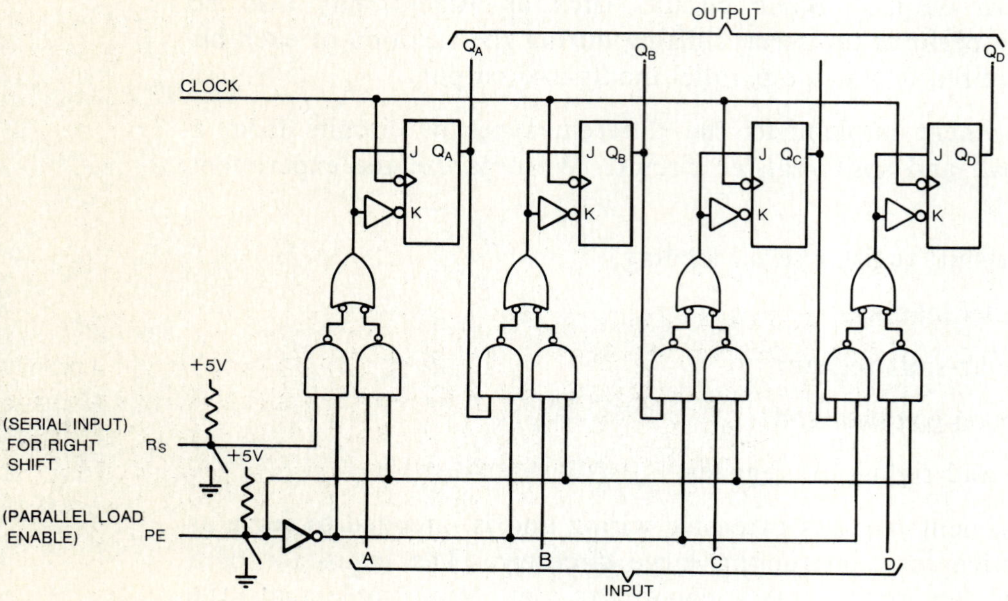

Figure 4-17. *The Shift-Right Register with Parallel Load*

Equipment:

A logic trainer.

Procedures:

4.4.1 Part A - Right-Shift Register Serial Loading

1. Connect the circuit shown in Figure 4-17. The circuit wiring diagram is shown in Figure 4-19 for reference.
2. Although not used in Part A, the parallel inputs A, B, C, and D should be connected to the data switches for the use in Part B.
3. Connect the RS signal to a data switch as well as the PE signal to a data switch.

Note that JK FFs are trailing-edge triggering in 7476 chips. When PE is high, the circuit performs parallel load. When PE is low, the circuit performs serial shift. If RS is high, the serial shift input is 1. If RS is low, the serial shift input is 0.

4. Connect the clock lead to a logic switch.

5. Connect the JK outputs (Q_A, Q_B, Q_C, and Q_D) to the four LED displays.

6. Initialize the circuit for serial operation by setting the PE switch in the LOW position.

7. The parallel input switches can be placed in any position as the parallel inputs are disabled.

8. Apply power to the circuit.

   Note: The state of the JK FFs is unknown when power is first applied. To "clear" the circuit, set the RS switch in the LOW position and apply enough clock pulses to turn off all LEDs.

9. Input the hex value A, 1010, into the shift register by changing the input RS switch and applying clock pulses.

10. On each clock pulse used to input data, observe and record the state of the display LEDs.

4.4.2 Part B - Parallel Loading

1. Initialize the circuit by setting the RS switch to the HIGH position and applying enough clock pulses to turn all LEDs ON. (It should only require three clock pulses.)

2. Change the operation of the circuit from serial to parallel by placing the PE switch in the HIGH position.

3. Input the hex number 5, 0101, into the shift register by setting the A through D switches and applying a clock pulse.

4. Observe and record the change of LED states in a table.

5. Place the circuit in the serial mode by placing the RS switch in the LOW position, and by applying clock pulses shift the 0101 out of the register. At each clock pulse observe and record the LED states.

6. Observe the difference between serial and parallel loading of the shift register.

4.4.3 Part C - Circular-Shift Register

1. The shift register illustrated in Figure 4-17 can be used to perform circular shifting by connecting the Q_D output back around to the RS. First, set PE to logical 1 and load binary input 0101 into the register. Then, reset PE to logical 0 and connect the circular-shift register as described. Let the circuit operate automatically from the 1-Hz clock pulse. Explain the difference between a shift register and circular-shift register.

4.4.4 Part D - Johnson Counter

1. Connect the Q_D' output of the FF as shown in Figure 4-17 to the RS. Assume the register initially contains all 0s. Record the possible register states (starting from 0000) if the circuit is operated from the 1-Hz clock pulse. Since the circular connection is made from the complement output of the rightmost FF, the register right-shifts its contents with each clock pulse, and at the same time, the complement value of the rightmost FF is transferred into the leftmost FF.

2. Turn off power to modify the circuit for implementing left-shift operations.

4.4.5 Part E - Left- and Right-Shift Register Circuit

1. Study Figures 4-17 and 4-18 noting the differences and similarities.

   Note: The serial input is used for both the shift-right and shift-left input. Shifting left requires connection of Q_B to A, Q_C to B, and Q_D to C. Analyze the reason for these connections so as to answer question 2 of the experiment.

2. Modify the circuit implemented from Figure 4-17 to reflect Figure 4-18.

3. Apply power to the circuit.

   Note: The state of the JK FFs is unknown when power is first applied.

4. Input the binary pattern 1010 by placing the Mode (MD) switch in the LOW position, and alternate placing the RS in the LOW position applying a clock pulse, then the RS in the HIGH position and applying a clock pulse. Record the state of LEDs.

5. Place both the RS and the MD switches in the HIGH position and apply clock pulses. Notice that you are operating a left-shift rather than a right-shift.
6. Record the state of the LEDs.
7. Repeat Steps 5 and 6 three times.

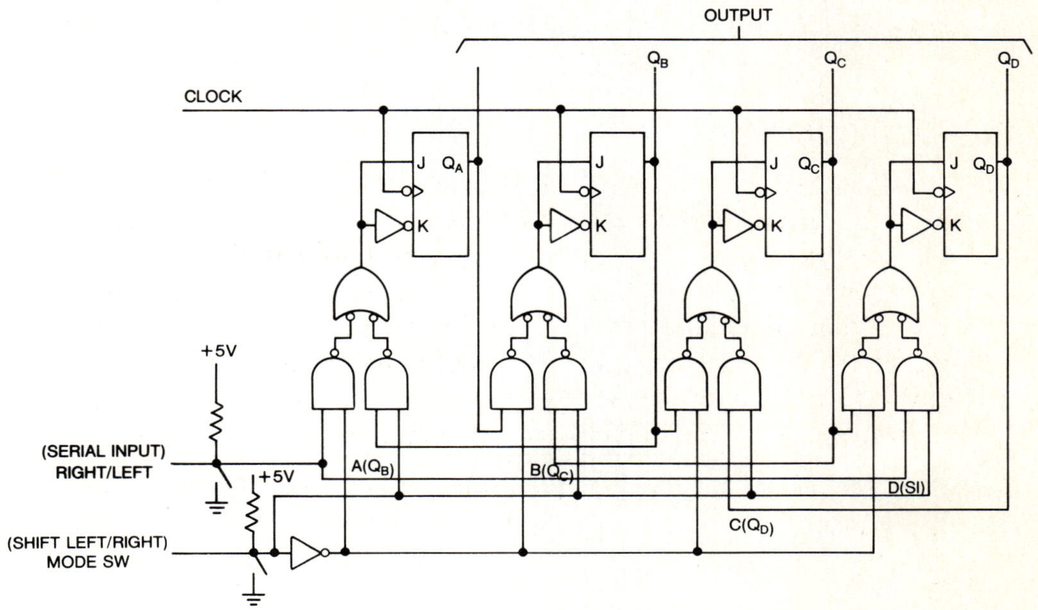

Figure 4-18. *The Shift-Right/Shift-Left Register*

Questions:

1. Describe how a shift-right/left circuit as shown in Figure 4-18 could be used to perform multiplication and division on binary numbers. Explain the necessary external inputs to the circuit.
2. Design a single 4-bit parallel load register without shifting functions. Explain the technique of converting the parallel load register to a shift-right or a shift-left register.

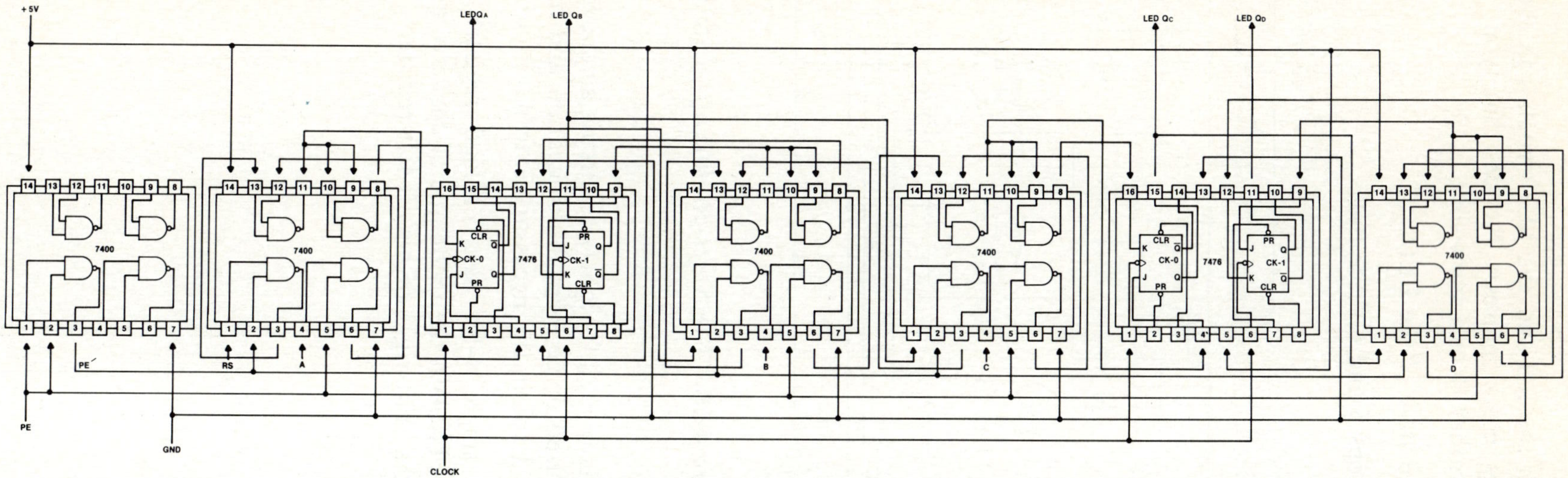

Figure 4-19. *Wiring Schematic of the Shift Register with Parallel Load*

4.5 Experiment 5 - Binary Coded Decimal (BCD) Counter and 7-Segment Display

Introduction:

All the electronic circuits studied in this book worked with two states or conditions and, therefore, could perform operations using only binary or 2-state arithmetic. Calculators, computer arithmetical circuits, and test equipment (for example, frequency counters) must perform decimal arithmetic. In general, components used in decimal arithmetic circuits are also binary devices; therefore, decimal numbers must be encoded into a binary representation. It requires four binary digits to represent a single decimal digit; each binary digit can be represented by a JK FF or a latch.

There are a variety of codes available to perform a decimal count sequence using binary numbers. A few common codes that can be found in the literature are the 8421 code, the Gray code, the Excess 3 code, the 2421 code, and the 5421 code. Figure 4-20 illustrates the counting sequence of a single decimal digit of the 8421 code used in this experiment. It is often necessary to encode a decimal number into a binary form, operate on it in binary form in the computer, and then decode the number back into decimal form.

The BCD counter counts from 0 to 9 and then back to 0 again. The counting sequence of a decimal number is very much part of our daily life. A 7-segment display is widely used for the display of a decimal digit in electronic watches, clocks, and appliances. The second portion of the experiment teaches the operation of a 7-segment LED display for displaying the counting sequence of the BCD counter. We will explain the principles of the 7448 7-segment decoder driver and the 7-segment display at the end of the procedure portion of the experiment.

The BCD counter circuit design is also shown in Appendix B as a synthesis example.

Objectives:

To study the behavior of the BCD counter circuit and to demonstrate the operation of the 7-segment decoder driver and 7-segment LED decimal display.

Materials:

1. Two 7476 dual JK FF chips.

2. Two 7400 quad 2-input NAND gates.
3. One 74154 4x16 decoder.
4. Extra LEDs.
5. One 7448 7-segment decoder driver.
6. One 7-segment display.

DECIMAL SYMBOL	8421 CODE	HARDWARE REPRESENTS ABCD
0	0000	0000
1	0001	1000
2	0010	0100
3	0011	1100
4	0100	0010
5	0101	1010
6	0110	0110
7	0111	1110
8	1000	0001
9	1001	1001

Figure 4-20. *8421 BCD Code*

Equipment:

A logic trainer.

Procedures:

1. Connect the circuit as shown in Figure 4-21 on the logic trainer. The circuit wiring diagram is shown in Figure 4-24 for reference. Reset the counter to zero.
2. Test the circuit by advancing the counter one pulse at a time and record the output after each clock pulse. Compare and verify the experimental results against the desired count sequence as indicated in Figure 4-20.
3. Connect your counter circuit to the 4x16 decoder chip (review the decoder in Experiment 2).
4. Connect the 7-segment display driver circuit as shown in Figures 4-22 and 4-23. Observe and record the counting sequence in the display.

The 7-segment display consists of seven LED segments arranged in a shape shown in Figure 4-23. The 7-segment decoder/driver chip 7448 is to be used with the 7-segment display; the 7448 chip pin layout, numerical designation, and function are provided in Figure D-8. The inputs to the 7448 driver come from one decimal digit in the BCD form.

The seven outputs, a through g, are meant to be connected to a 7-segment display. If one of the outputs (a through g) is ON, the LED segment connected to the output illuminates. For example, if inputs DCBA are 0001, which is the BCD code for 1, only segments b and c glow. They form a single rightmost vertical line and give the appearance of the numeral 1. Figure 4-23 shows how a 7448 drives a 7-segment display.

The LT and RBI inputs can be used to implement remote blanking. Remote blanking is a feature of decoder/drivers that allows them to blank out leading 0s for multiple decimal digits. We will not investigate this feature of the 7448 in this book.

Questions:

1. The circuit in Figure 4-21 is a single decimal BCD digit up counter. On paper, design a 2-digit 8421-code up counter. The circuit in Figure 4-21 is technically called an 8421 decade. For correct operation, both decade stages should operate synchronously (on the same clock pulse).
2. Explain the relationship between the decade stages in a multistage 8421 counter.

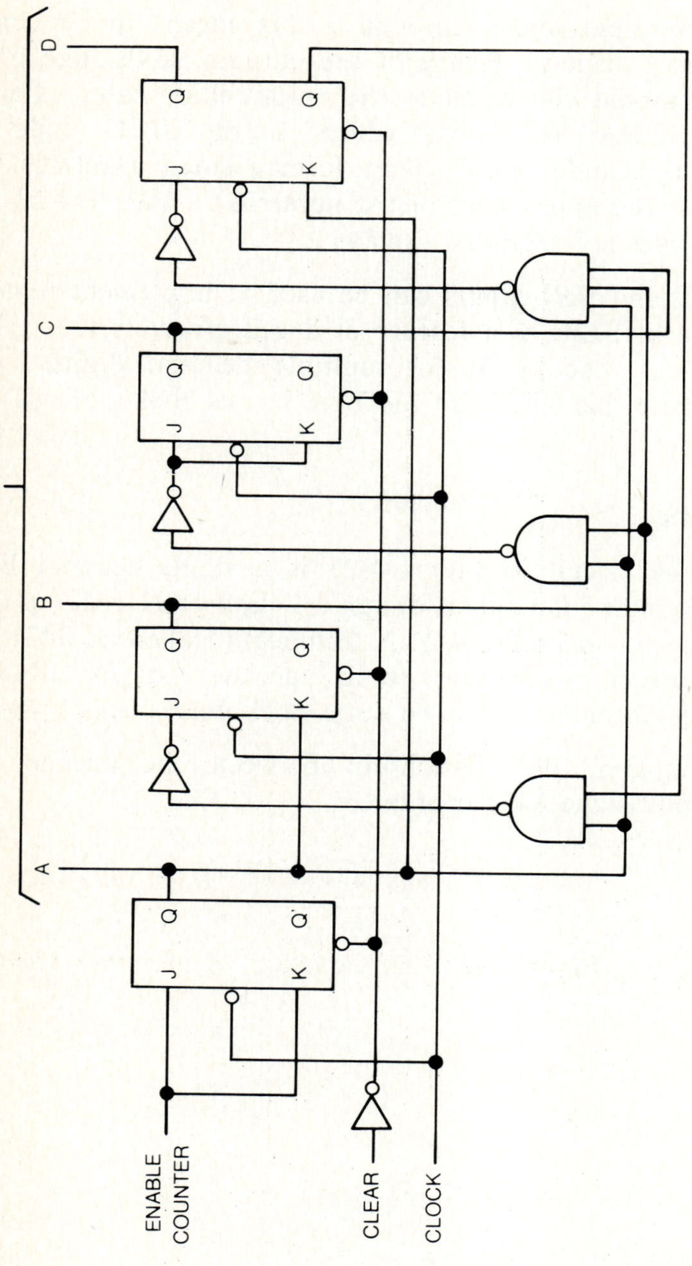

Figure 4-21. *8421 BCD Counter*

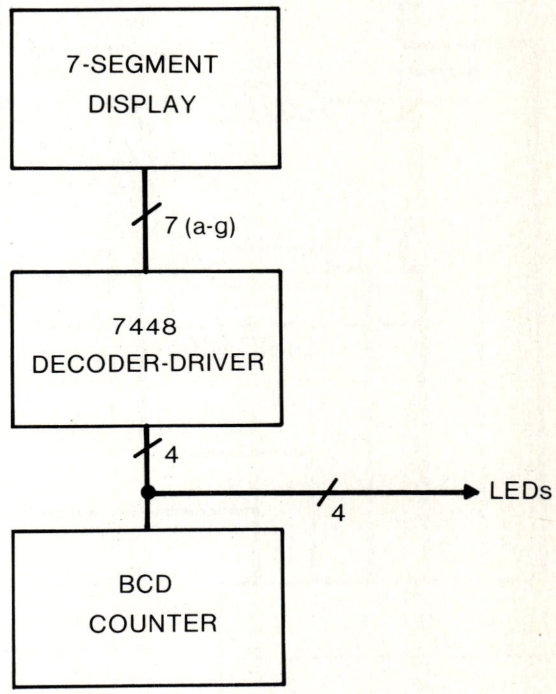

Figure 4-22. *Block Diagram of Experiment 5*

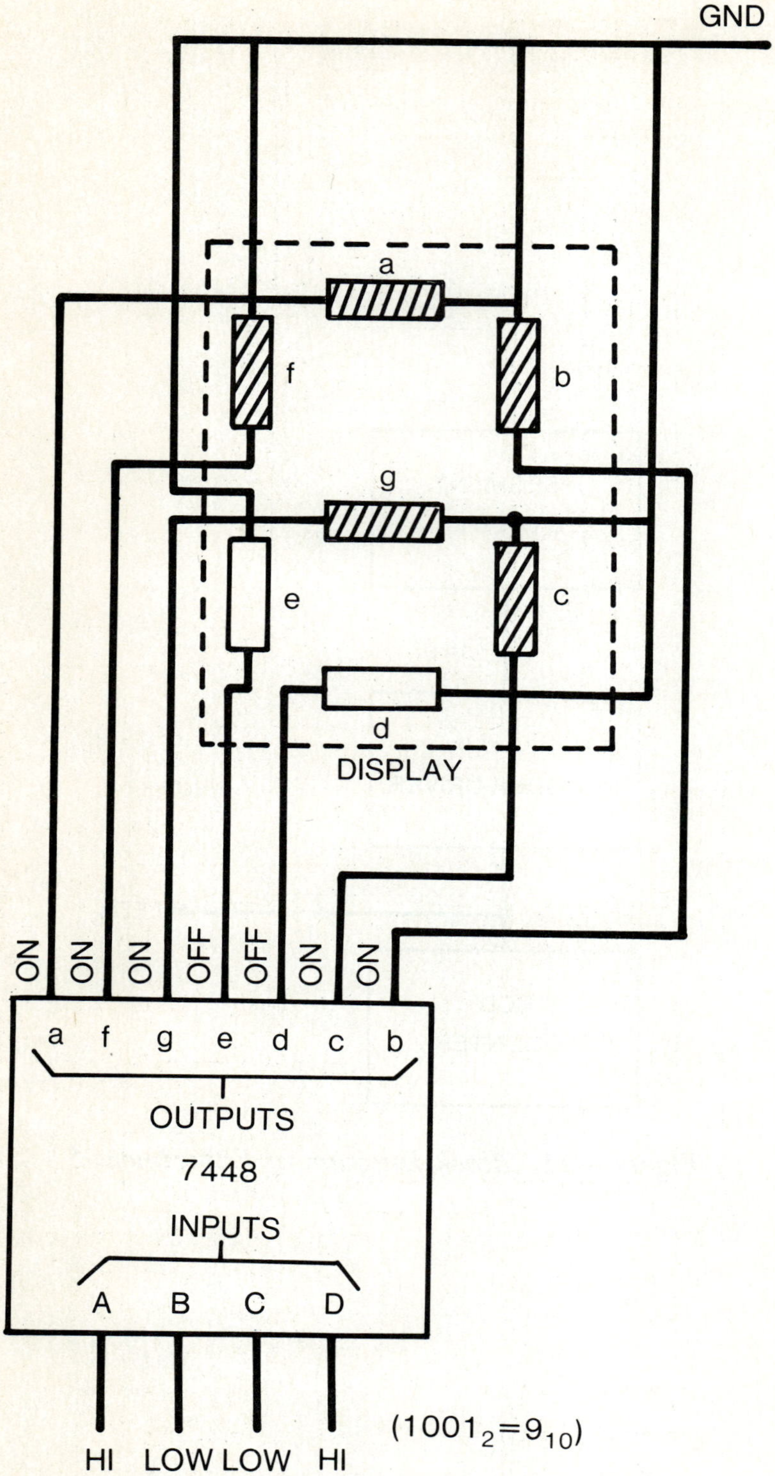

Figure 4-23. *A 7448 Driving a 7-Segment Display*

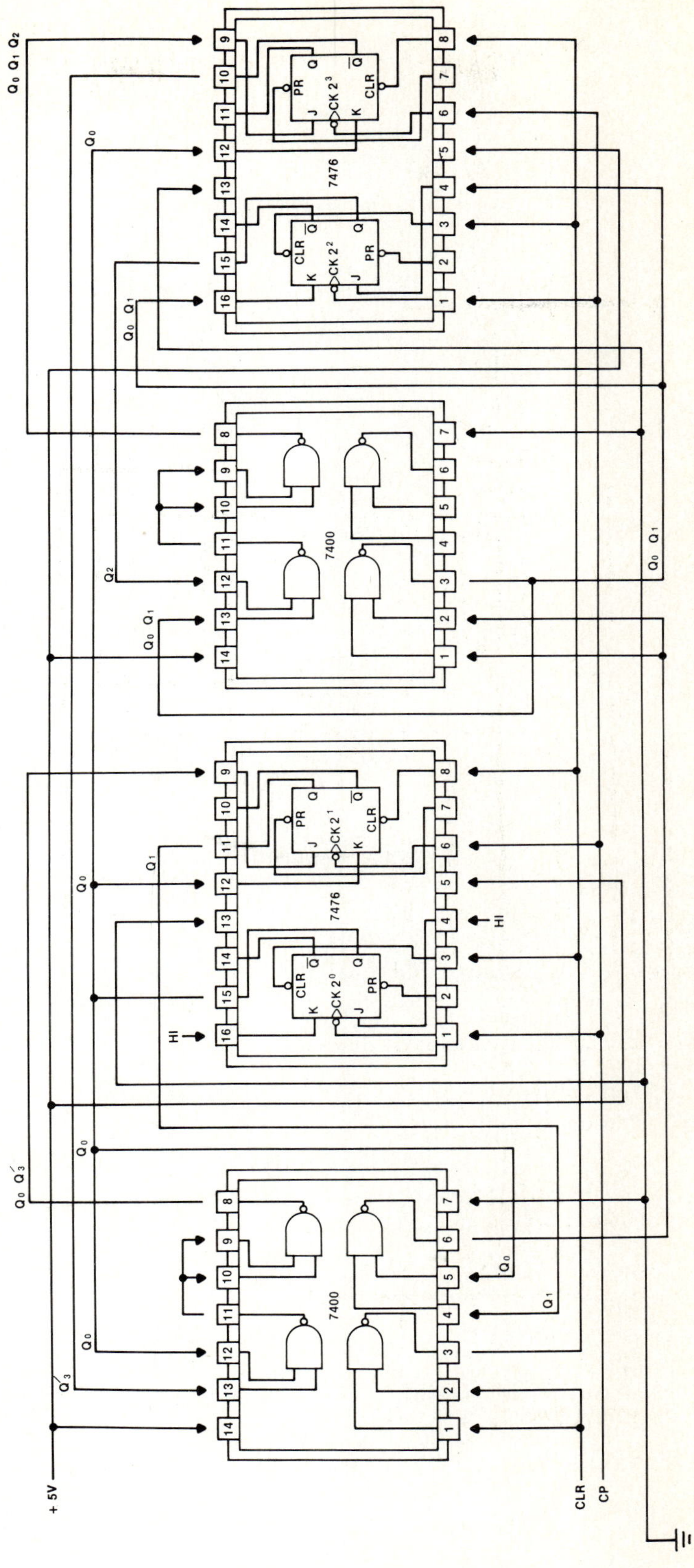

Figure 4-24. *Wiring Schematic of the 8421 BCD Counter*

4.6 Experiment 6 - Random Access Memory

Introduction:

The data and instructions of a computer program have to reside in the memory during its execution. The memory can be either volatile or nonvolatile. A memory unit is volatile if its contents are lost during a power-off period. This type of memory is often called random access memory or RAM. In order to permanently store the information that is on RAM, magnetic tapes or disks should be used. The nonvolatile type of IC memory is called read only memory or ROM. The contents of the ROM stay intact even when the IC's power is turned off. ROM is often used to store programs that affect the digital system's operation such as the operating system of a computer. RAM is used to store any type of program.

A standard memory chip now has a capacity of 64K bits. The capacity of memory that can be placed on one chip is increasing rapidly. The new generation memory chip can contain 256K bit memory. About 30 pages of text, at 250 words per page, can be stored on a single 256K memory chip. In this experiment, we will study the detailed operation of RAM only. A 7489 RAM chip has 16 unique address units called words. Each word has 4 bits. The 7489 is a 64-bit RAM whose pin layout and function table are shown in Figure D-6. The input to the 7489 consists of:

- 4 data input bits,
- 4 data output bits,
- 4 select bits (to select 1 of 16 words), and
- two control lines: Chip Enable (CE) and Write Enable (WE). The CE of the RAM is also called Memory Enable (ME) in some books.

As the function table shows, this allows the designer four options:

- READ,
- WRITE,
- invert data, or
- do nothing.

The *Do Nothing* option is very important because it de-selects the IC. Since the 7489 has open collector outputs, another 7489 can be paralleled with it and selected. This increases the size of the memory by 16 words for each additional IC.

Note that we only use four 2-bit words from the 7489 chip in the experiment. This is required to simplify the wiring of the implementation. The RAM is both readable and writable. To read, an address is placed in a Memory Address Register (MAR) and a READ signal is given. The contents of the reading address are loaded in a Memory Buffer Register (MBR). To write, the writing address is placed in an MAR, the data is placed in an MBR, and a WRITE signal is given.

The block diagram of the RAM circuits is shown in Figure 4-25. It shows the external connections of the 7489 chip. The actual implementation of the RAM circuit for experimentation is also shown in Figure 4-25.

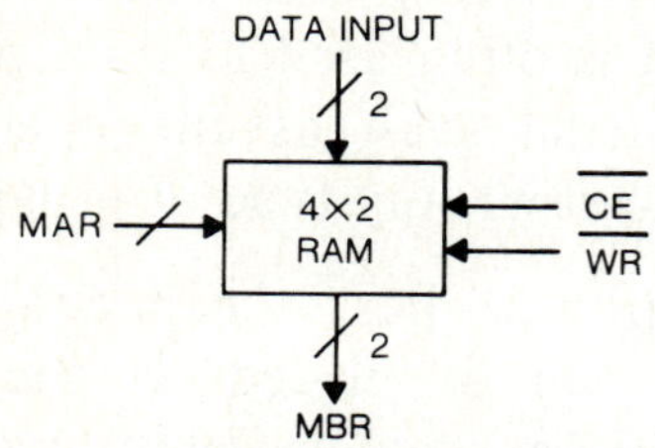

(a) Block Diagram

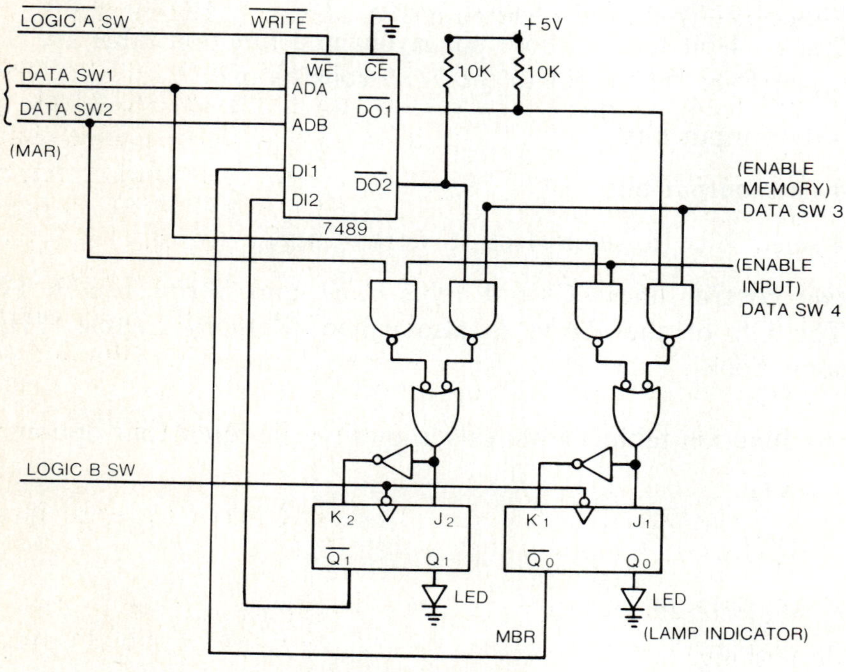

(b) Circuit Diagram

Figure 4-25. *RAM Memory (4-Word by 2-Bit Memory)*

Objectives:

To examine the behavior of a simple memory circuit.

Materials:

1. Two 7400 quad 2-input NAND gates.
2. One 7476 dual JK FF chip.
3. One 7489 16x4 RAM memory chip.
4. Two 10K-ohm resistors as shown in Figure 4-8. These resistors are required since the 7489 outputs are open collectors.

Equipment:

A logic trainer.

Procedures:

1. Connect the circuit as shown in the circuit diagram of Figure 4-25. Connect the Q side of the MBR JK FFs to indicator LEDs 3 and 4. The circuit wiring diagram is shown in Figure 4-27 for reference.
2. Connect the memory outputs to indicator LEDs 1 and 2.
3. Connect the complement of logic switch A to the memory WE lead. Logic switch B is used for MBR clock pulse activation. Figure 4-26 illustrates an action sequence chart for the simple RAM memory circuit. The first column designates the particular action (for example, load MBR, write into memory, or read from memory). The second column shows the meanings of the values associated with data switches 1 and 2 for the associated action. Columns 3 and 4 designate values for select memory and select input switches (data switches 3 and 4, respectively) and the remaining two columns indicate whether a clock pulse or a WE pulse is required to cause the indicated action.
4. Write a detailed report on the steps taken to write into and read from the 7489 memory chip.
5. Perform the following steps and prepare a table to record the content of address, input, MBR, and memory output along with the values associated with logic switches A and B.

    - Write into memory all 0s, then read memory.
    - Write into memory all 1s, then read memory.
    - Write the 1's complement of the address into memory, then read memory.

6. Implementation Notes:

   For the purpose of simplicity, the 1's complement of the data is actually written into memory. For example, when binary value 10 is written into memory, the value 01 is actually written into the memory location. This special arrangement can read back data correctly since the memory outputs are complemented.

   The MAR is connected to the data switches to save an FF chip. The MAR is also used to serve as an external input to the MBR.

Questions:

A memory unit has a destructive read property if the content of a word is destroyed after a memory read.

1. Is this memory unit volatile? Does it have a destructive read property?
2. The 7489 RAM consists of open collector outputs. When the chip is not enabled (CE'=1), the output will be all 1s. The block diagram of the 7489 RAM chip is similar to the block diagram shown in Figure 4-25. The memory size of the 7489 is 16x4. Experiment 6 used only a 4x2 portion of the memory.
   a. Show the external connections to two 7489 chips that will produce a 32x4 RAM.
   b. Show the external connections to two 7489 chips that will produce a 16x8 RAM.
   c. Show the external connections to eight 7489 chips that will produce a 64x8 RAM; use a 2x4 decoder constructed with NAND gates. (Hint: The CE can be used as a chip select for system expansion.)

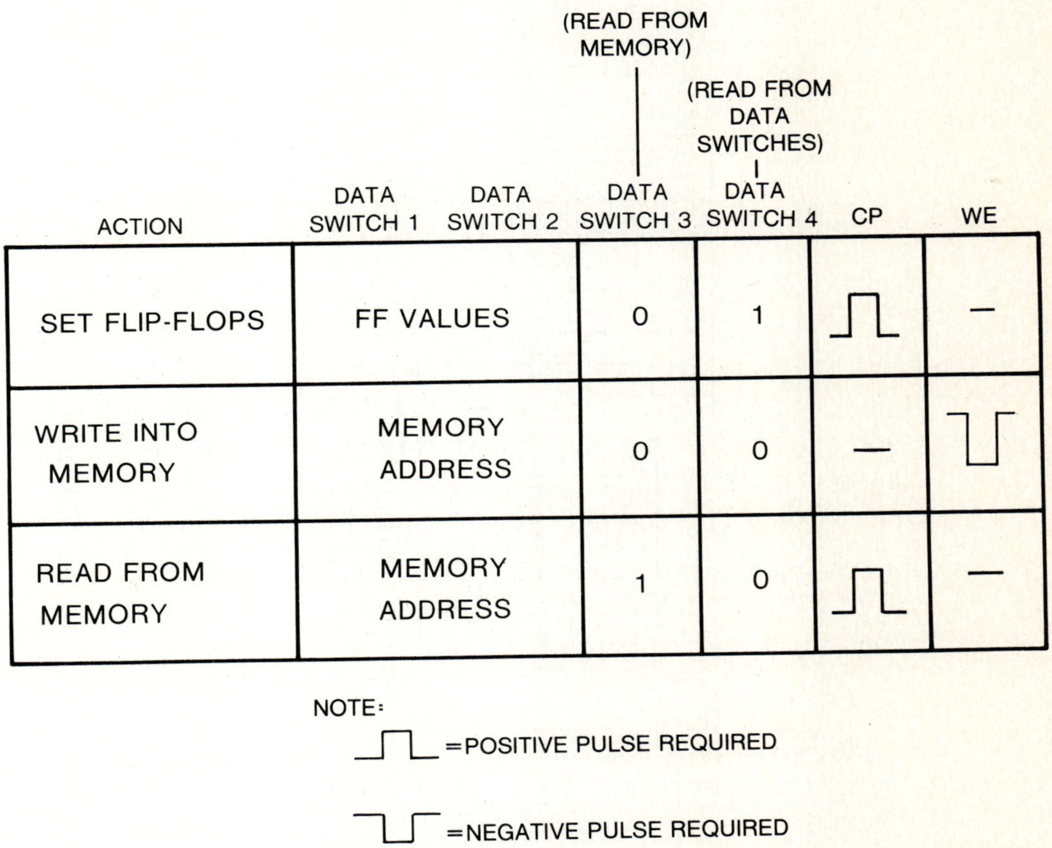

Figure 4-26. *Action Sequence Chart for RAM Memory Experiment*

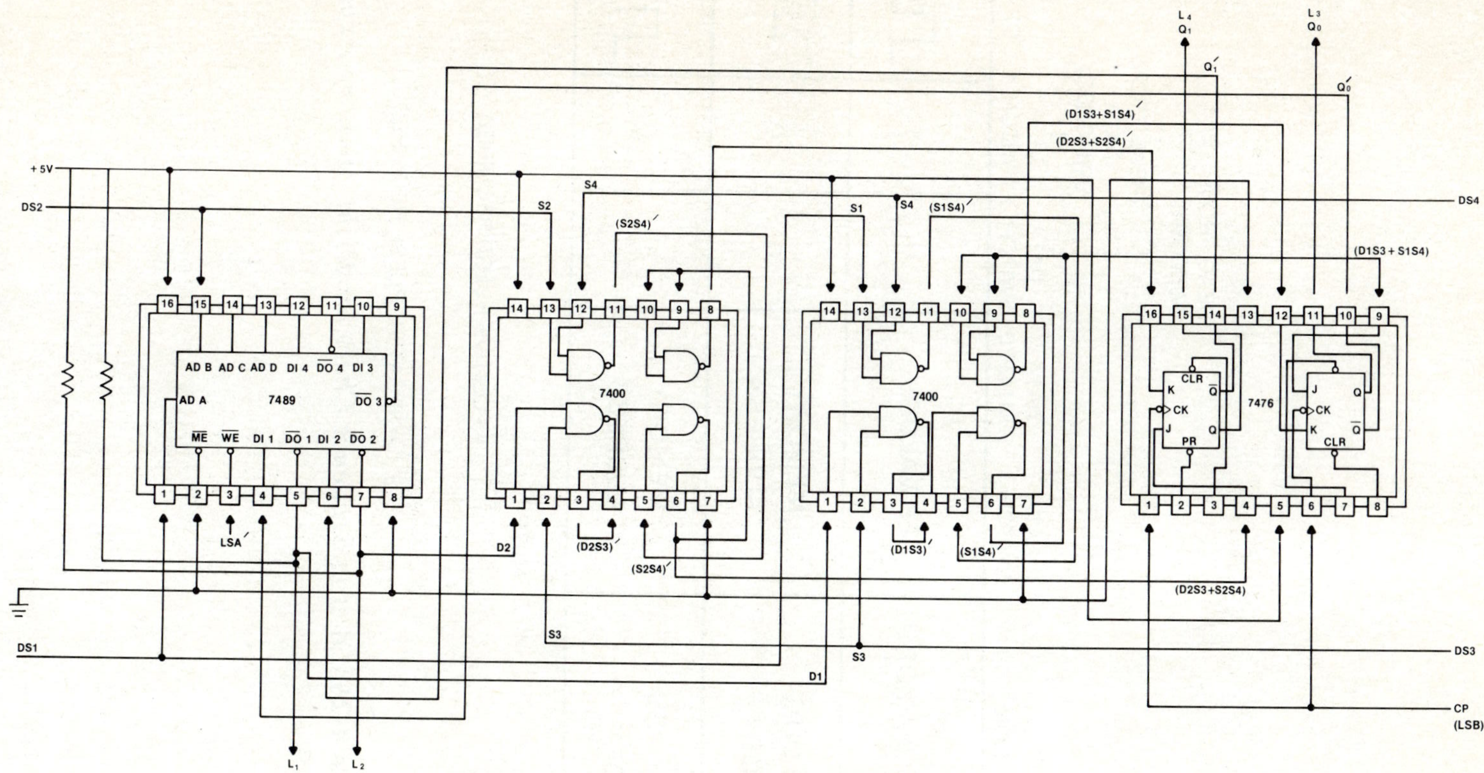

Figure 4-27. *Wiring Schematic for RAM Memory*

5

Digital System Experiments

In the instructional experiment, we have learned to build high-level digital functions such as the parity detector, shift register, counter, and memory circuits from basic building blocks such as exclusive-OR gates, NAND gates, and FFs. We can represent high-level digital functions by block diagrams such as the one used in the RAM circuit diagram in Experiment 6 so that the functions can be treated as new building blocks for building more complicated designs. The circuits we have studied in the instructional experiments provide adequate building blocks for us to build all of the circuitry described in the rest of the book. The building-block process is similar to generating various macros in an assembly language for use in an assembly language program.

The digital system experiments are intended to provide you with some hands-on design and implementation experience by using the circuit construction skill learned from the instructional experiments. It would be helpful if you have some knowledge of computer architecture and digital systems for completing your design. The experiments have been implemented by the students of the computer architecture courses at Illinois Institute of Technology during the past 3 years. We have intentionally left out some of the design steps so that you can design your own circuits and then verify their correctness by experimentation. You may need to refer to the synthesis and RT language sections of the Appendixes to understand the logical steps and notations used in the digital system design of this book.

There are six experiments in this chapter:

Exp. 7 — Up-Down Counter and Word-Time Control Signal Generation Circuit

Exp. 8 — Traffic Light Controller

Exp. 9 — Memory Stack Circuit

Exp. 10 — Hardware Communication Queue

Exp. 11 — Communication Protocol Machine

Exp. 12 — Programmed Digital Controller

Experiments 7, 8, and 12 are recommended for implementation in a supervised laboratory. Experiments 9, 10, and 11 convey similar concepts and the instructor may decide to implement one of them and leave the others as paper-and-pencil exercises. Experiments 7 and 8 take 1 to 2 weeks to complete. Experiments 9 through 12 take 2 to 3 weeks to complete.

Due to the amount of work required in the implementation of the experiments, it is essential that you understand the experiments before the circuit construction. For a supervised laboratory course, you should use the laboratory session only for constructing and debugging circuits and for experimental verification of the design.

Four optional design exercises are included in Sections 5.7 through 5.10. Exercise 1 is the design of an odd-even counter. It can be used to replace Experiment 7. Exercise 2 is the design of a furnace temperature controller. It can be used to replace Experiment 8. Exercise 3 is the design of a multiplication circuit. Exercise 4 is the design of an asynchronous communication circuit. It can be used to replace Experiment 11.

If you are interested in performing more challenging experiments, Reference 4 of Logic Design in the Bibliography has some more design projects for your further study. This book also discusses generalized concepts in sequential circuits and tuning machines.

5.1 Experiment 7 - Up-Down Counter and Word-Time Control Signal Generation Circuit

Introduction:

With an understanding of the seven design steps required in the circuit synthesis section in Appendix B, you can design a counter circuit in Part A of the experiment that can count either upward or downward by control selections. The up-down counter is then used as a building-block component to the larger word-time control signal generator circuit. The word-time control signal lasts for several clock periods that have the same count as the number of bits in a word. Part B of the experiment implements a control signal for a 4-bit word. The signal can be used to control serial operations of the word such as addition, shifting, etc.

Objectives:

To design and implement a word-time generation circuit using a 2-bit up-down counter circuit.

Materials:

1. 7400 quad 2-input NAND gates.
2. 7420 dual 4-input NAND gates.
3. 7476 dual JK FFs.
4. One 7495 shift register (optional).

Equipment:

A logic trainer.

Procedures:

5.1.1 Part A - Up-Down Counter

1. Design and implement a 2-bit up-down counter circuit with an enable input terminal and separate up and down controls for both the up and down counts.
2. Be sure to address the following areas in the laboratory report:

Objectives
Materials
Equipment
Circuit Synthesis
Procedure and Observations
State Table for the Present State, Inputs, Next State, and Outputs

5.1.2 Part B - Word-Time Control Signal Generation Circuit

1. Using the up-down counter circuit implemented from Part A, design and implement a word-time generation circuit as shown in Figure 5-1. The up and down inputs to the circuit should change on the trailing edge of the STOP signal. Both the counter and the word-time control FF should be initialized to the LOW state.
2. The timing diagram is suggested in Figure 5-2.
3. Be sure to include the discussion as specified in Part A as well as provide a complete timing diagram for the word-time generation circuit.
4. The start signal, S, can be connected to a pulser switch.
5. (Optional) If the 7475 shift register is available, the word-time generator circuit may be used to control the circular shifting of the entire register with one start signal.

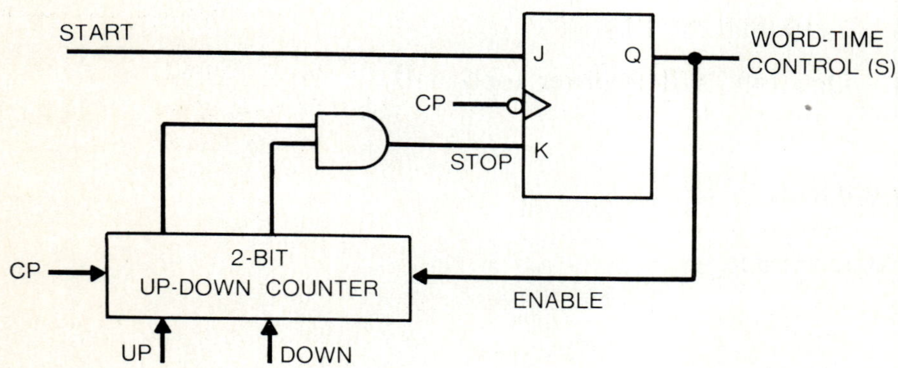

Figure 5-1. *Word-Time Control Signal Generation Circuit*

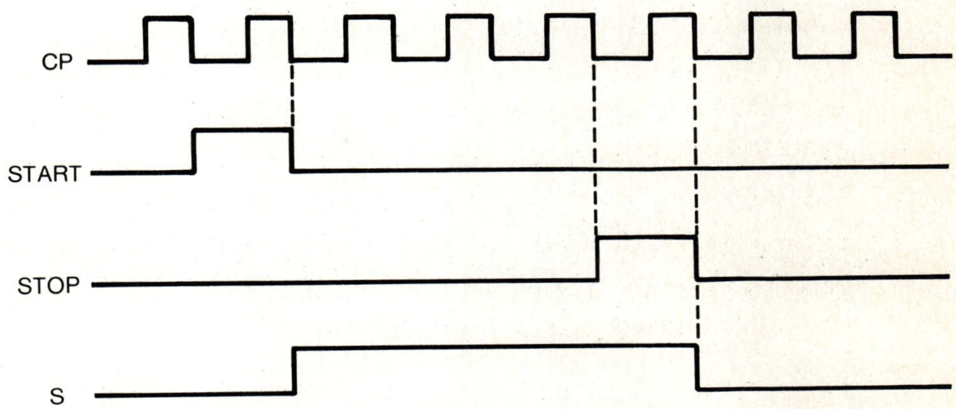

Figure 5-2. *Word-Time Control Signal Generation Timing Diagram*

5.2 Experiment 8 - Traffic Light Controller

Introduction:

There are many applications for controller circuits; one of these is a highway intersection traffic light. Pressure activated sensors are usually embedded into the roadway to determine the sequence of the light changes to optimize the flow of traffic. In this experiment, the duration of a vehicle waiting at a stoplight is measured by counting the number of clock pulses during the time the vehicle was sensed waiting at a red light. This experiment is a simple traffic light problem for a 4-way stoplight, one in each direction. The traffic light controller is designed with requirements that will allow you to implement it with just a few chips. We also start to use the RT language, described in Appendix C, to specify the design.

Objectives:

1. to expand your knowledge of sequential circuits,
2. to develop an understanding of possible uses of sequential circuits, and
3. to develop an appreciation of hardware applications to our everyday life.

Materials:

1. 7400 quad 2-input NAND gates.
2. 7420 dual 4-input NAND gates.
3. 7476 dual JK FF chips.

Equipment:

1. A logic trainer.

Procedures:

1. Design and implement a traffic control system for an intersection of two highways. (Refer to Figure 5-3):

   a. Basic Components:

      – a North/South traffic light, N/S_Light (green = LED off, red = LED on),

- an East/West traffic light, E/W_Light (green = LED off, red = LED on),
- a light control FF for the East/West light,
- a light control FF for the North/South light.

The light control can be implemented by one light control FF, since the N/S_Light is always the complement of the E/W_Light.

When a car has been detected waiting at a red light, a counter circuit is enabled to provide a delay of four clock pulses followed by a light change. The counter shown in the circuit diagram in Figure 5-4 is a ripple counter.

The scanner and counter circuits are driven by four sensors at the four corners that detect vehicles waiting for the light to change. The four sensors are North_Sensor, South_Sensor, East_Sensor, and West_Sensor.

b. Traffic Light Operation:

- To change a traffic light, a vehicle usually must wait at a red light for at least three clock pulses; the fourth clock pulse changes the light. If a vehicle stops and makes a right turn, the counting stops. To cause the counting to continue, another vehicle must stop at the red light.

(1) Initialize Circuit

INITIALIZE: counter ← 0,
E/W_Light ← red,
N/S_Light ← green

(2) Change Light

((East_Sensor + West_Sensor)·E/W_Light + (North_Sensor + South_Sensor)·N/S_Light)·clock:
counter ← counter + 1,
if (counter = 3) then
E/W_Light ← (E/W_Light)',
(N/S_Light) ← (N/S_Light)'

Note: N/S and E/W lights are inverse of each other.

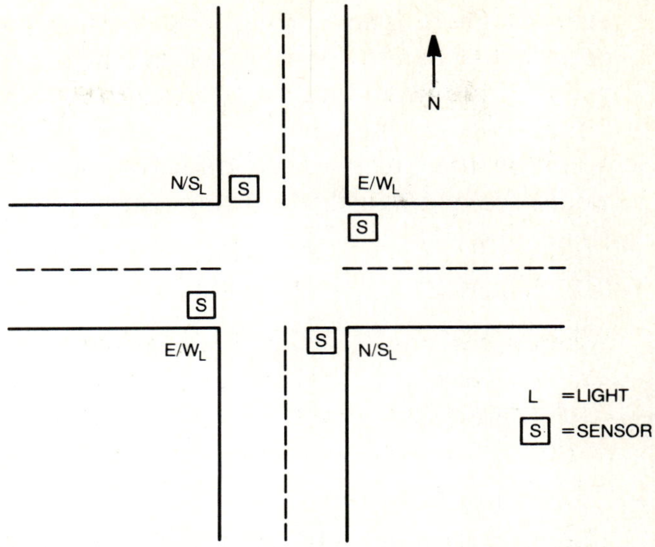

Figure 5-3. *A Traffic-Light Intersection Diagram*

2. Implementation Notes:

   – A data switch may be used to implement each sensor. A sensor is on whenever a vehicle is detected due to the vehicle's pressure on the sensor.

3. Testing Procedures:

   – To test this circuit, first initialize the circuit for the North/South direction. All sensors (data switches) should be in the NOT state. Set the North sensor; there should be no change in the light status. Return the North sensor to the NOT state and set the East sensor. The lights should change. Leaving the East sensor set, set the South sensor. The lights should now cycle. Put the South sensor then in the NOT state. The lights should indicate the East/West green condition. Now set the North sensor and again the lights should cycle.

Questions:

1. A sensor can also be used to detect traffic flow (traffic movement). Traffic flow is generally measured by scanning a sensor and accumulating a count over a known period of time. Should the sensor be placed near or far from the intersection and why?

2. How can the traffic light circuit be modified to include a

yellow light? The current circuit times for four clock periods to change from red to green/green to red. The yellow light would occur after three clock periods and remain on for one clock period. The yellow light would occur on the green to red directions and the green light would be extinguished during the yellow light period. Remember that the light must not stay at yellow for more than one clock time.

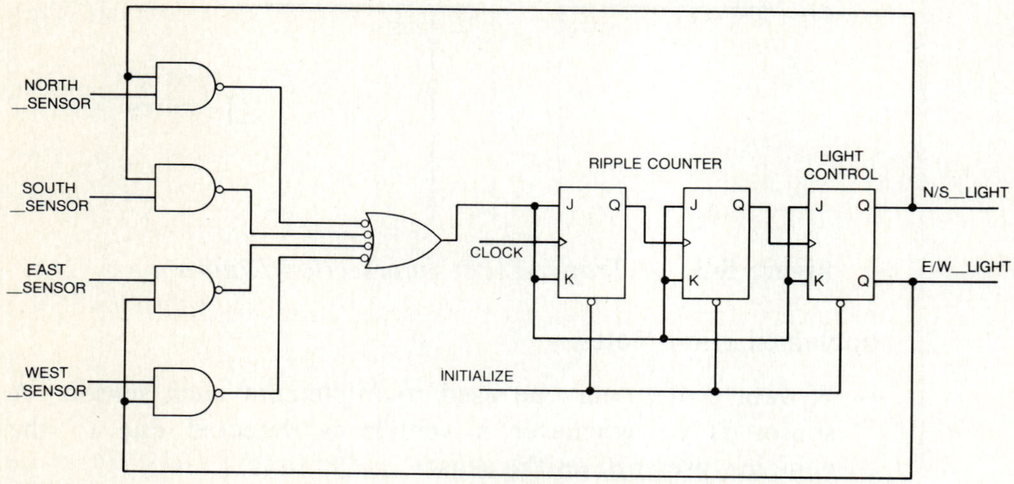

Figure 5-4. *The Circuit Diagram of the Traffic Light Controller*

5.3 Experiment 9 - Memory Stack Circuit

Introduction:

A stack is a last-in/first-out data structure. It is an extremely useful data structure in handling subroutine linkage and in writing a compiler. A stack can be implemented by either software or hardware. The major motivation of implementing the hardware stack is to gain the speed in stack-related operations. The hardware stack may be implemented either by RAM or registers. This experiment implements a memory stack by hardware.

Objectives:

1. to use the knowledge learned to design a simple stack,
2. to understand timing sequences for micro-operations, and
3. to develop an appreciation of tradeoffs between hardware and software.

Materials:

1. 7400 quad 2-input NAND gates.
2. 7420 dual 4-input NAND gates.
3. 7475 4-bit bistable latches.
4. 7476 dual JK FF chips.
5. One 7489 16x4 RAM memory chip.
6. 10K-ohm resistors.
7. Extra data switches and LEDs may be required.

Equipment:

1. A logic trainer.
2. An extended breadboard. (Be sure to ground the two boards together.)

Procedures:

1. Design and implement a memory stack with the following specifications (refer to Figure 5-5):

a. Basic Components:

- an eight 4-bit word stack memory, M,
- a 3-bit Stack Pointer Register (SPR), which is an up-down counter and points to the top of the stack, can be used to store the next item pushed into the stack,
- an MBR register comprising the 7475 chip; outputs from the FF are connected to LEDs,
- a full indicator (JK FF),
- an empty indicator (JK FF), and
- initially, SPR points to the bottom of the stack (location 0) and the stack is empty. The stack becomes full when an item is pushed into the top of the stack (location 7).

b. Stack Operations:

- There are three operations associated with the stack: PUSH, POP, and CLEAR. These operations are defined by the following micro-operations:

PUSH

$PUSH \cdot T_1$: $SPR \leftarrow SPR + 1$

$PUSH \cdot T_2$: $M \leftarrow$ (Data Switches), $EMPTY \leftarrow 0$, if $(SPR = 0)$ then $(FULL \leftarrow 1)$

POP

$POP \cdot T_1$: $MBR \leftarrow M$, $SPR \leftarrow SPR - 1$, $FULL \leftarrow 0$

$POP \cdot T_2$: if $(SPR = 0)$ then $(EMPTY \leftarrow 1)$

CLEAR

$CLEAR$: $SPR \leftarrow 0$, $EMPTY \leftarrow 1$, $FULL \leftarrow 0$

A PUSH operation is used when a new item is inserted onto the top of the stack. A POP operation is used when the top item on the stack is removed or deleted. The CLEAR operation is used to initialize or reset the stack to empty. Note that both PUSH and POP require timing signals T_1 and T_2 to control the sequence of operations to be performed. The circuit layout diagram of the stack is suggested in Figure 5-6.

From the RT language description, a table of micro-operation by registers can be derived, as shown in Table 5-1 of Experiment 12. This table is helpful in designing the pieces of the circuits involving each register. We leave the design of the complete circuit for the stack from the table of micro-operations as an exercise for you.

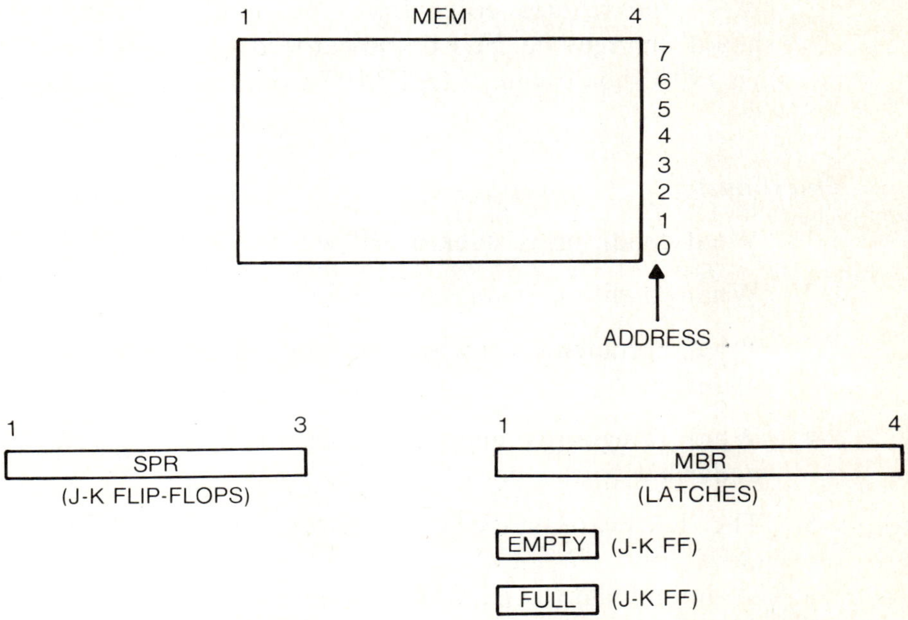

Figure 5-5. *The Hardware Stack Basic Component Diagram*

c. Implementation Notes:

- The 7489 RAM memory outputs require 10K-ohm pull-up resistors.
- Data into memory from the data switches is NOT data. NOT data can be considered to be the inverted data switches or the complement data represented by the data switches. The reason for the inversion was explained in the RAM experiment.
- The control unit of the machine is simulated by pulsers and data switches in order to generate the T_1, T_2, PUSH, and POP signals.
- SPR is a simplified up-down counter.
- POP and PUSH are two separate data switches and will never be a "1" at the same time.

d. Testing Procedures:

In order to demonstrate the stack operation, first CLEAR it. Then PUSH the value 0001 to 1000 onto the stack. The last PUSH should cause the FULL indicator to go ON. Then change the stack to the POP operation and read the values from the stack. Make sure the values pop out in the reverse order from the input. The first POP should change the FULL indicator back to OFF and the last POP should cause the EMPTY indicator to go ON.

Questions:

1. What condition is stack overflow?
2. What condition is stack underflow?
3. What problems are caused by stack overflow and underflow?
4. What safeguards may be added to detect overflow and underflow in the stack?
5. The hardware stack can perform PUSH/POP in one clock time. What changes must be made to modify and to implement a hardware register stack instead of a memory stack?
6. Are the memory locations in this organization being used effectively (in particular, location 0)? What design and implementation considerations must be made in this organization for a 4-bit SPR?
7. An excellent way of handling subroutine linkage in computer programs is 1) to push the return address onto a stack when a subroutine call is encountered, and 2) to pop the return address off the stack when a subroutine return is encountered. Explain the reason that the correct return address is always at the top of the stack under nesting or recursive subroutine calls.

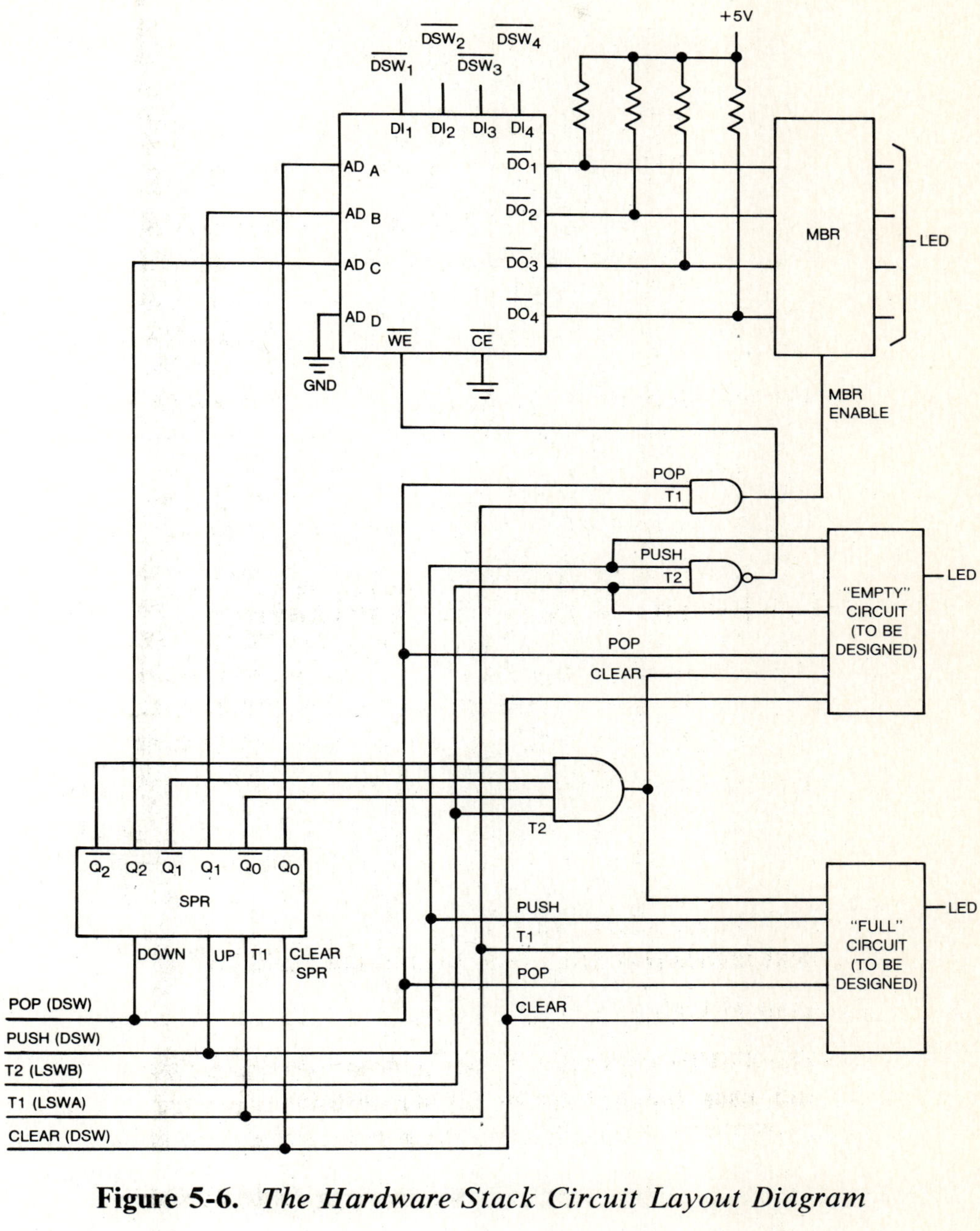

Figure 5-6. *The Hardware Stack Circuit Layout Diagram*

5.4 Experiment 10 - Hardware Communication Queue

Introduction:

When two microprocessors in a multiprocessor system must communicate with each other, a First In/First Out (FIFO) memory circuit is often used as a "mailbox" for sending and receiving messages (data) between the microprocessors. The FIFO data structure is very useful in handling a sequence of requests by the first-come/first-served method. Whenever fairness is a requirement, a pure FIFO queue or a priority-based queue is preferred to a LIFO stack. Using this method, one microprocessor can input data into the FIFO memory, and the other microprocessor can retrieve the data from the FIFO memory, in the same sequence. In a full-duplex system, two FIFO circuits can be used, allowing simultaneous communication between microprocessors. Each microprocessor would have one input FIFO memory and one output FIFO memory.

The FIFO memory implemented in Experiments 10 and 11 is less than a classical FIFO queue. A FIFO typically has separate address registers for input and output, and the value of the two must be compared to determine if the buffer is available for reading or writing.

Communication protocols are a set of rules that can be used to regulate the interactions between communicating entities so that they can exchange information successfully. The handling of the data passed between microprocessors must also conform to established protocol rules. The protocol for this experiment simply requires 16 bytes of data to be sent between microprocessors, and a write operation into the FIFO memory has to be followed by a read operation. Experiment 11 will handle more sophisticated protocols such as error checking and recovery.

Objectives:

1. to expand your knowledge of memory design, and
2. to develop an understanding of communication required between microprocessors in a multiprocessor configuration.

Materials:

1. 7400 quad 2-input NAND gates.
2. 7420 dual 4-input NAND gates.
3. 7476 dual JK FF chips.

4. One 7489 16x4 RAM memory chip.
5. 10K-ohm resistors.

Equipment:

1. A logic trainer.
2. An extended breadboard. (Be sure to ground the two boards together.)

Procedures:

1. Design and implement a FIFO memory with the following specifications. (Refer to Figure 5-7):
   a. Basic Components:
      - a 16x2 FIFO memory, M,
      - a 4-bit FIFO Address Register (FAR),
      - an MBR register, and
      - a READ/WRITE indicator (a single JK FF is used to implement these two separate control signals).
   b. FIFO Operations:
      - There are three operations associated with the FIFO memory: WRITE, READ, and CLEAR. These operations are defined by the following micro-operations:

WRITE

$WRITE \cdot T_1$: $M \leftarrow$ (Data Switches)

$WRITE \cdot T_2$: $FAR \leftarrow FAR + 1$,
if ($FAR = 15$) then ($WRITE \leftarrow 0$,
$READ \leftarrow 1$)

READ

$READ \cdot T_1$: $MBR \leftarrow M$

$READ \cdot T_2$: $FAR \leftarrow FAR + 1$,
if ($FAR = 15$) then ($READ \leftarrow 0$,
$WRITE \leftarrow 1$)

CLEAR

CLEAR: *FAR* ← 0, *WRITE* ← 1, *READ* ← 0, *MBR* ← 0

The WRITE operation is used to input data into the FIFO. The protocol between the two microprocessors requires 16 data bytes to be transferred at a time. Therefore, all 16 bytes must be written into the FIFO before data can be removed. A READ operation is then used to retrieve the data from the FIFO. Again, the protocol requires all 16 bytes to be removed before new data can be inputted to the FIFO. The CLEAR operation is used to initialize or reset the FIFO control to the WRITE mode and the FIFO address register to 0. Note that both WRITE and READ are controlled by the timing signals T_1 and T_2 from the same timing generator to control the sequence of operations to be performed. The block diagram of the circuit is suggested in Figure 5-8.

The READ/WRITE FF serves as a flag (or semaphore used in the operating system process synchronization) for the synchronization between the two microprocessors.

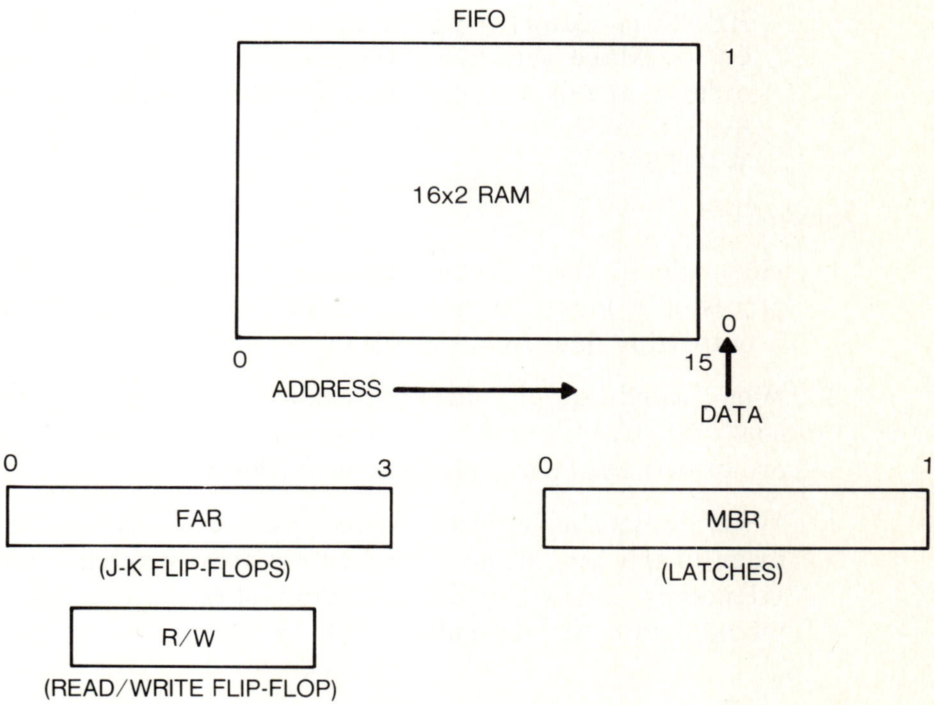

Figure 5-7. *The Basic Component Diagram of the Hardware Communication Queue*

- Implementation Notes:
  - The 7489 RAM memory outputs require 10K-ohm

pull-up resistors.

- Data into the memory from the data switches is inverted data. The reason for the inversion was explained in the RAM experiment.
- The control unit of the machine is simulated by the pulser and data switches in order to generate the T_1, T_2, and CLEAR signals.
- FAR is an UP counter.
- READ and WRITE are two separate control signals and will never be a "1" at the same time. They are implemented by a single FF.
- CLEAR is a separate switch.

- Testing Procedures:
- In order to demonstrate the FIFO operation, first CLEAR the circuit. Then WRITE the values 00,01,10,11...00,01,10,11 until the FIFO is filled. After writing the 16th byte, the control should change the READ indicator to ON and the WRITE indicator to OFF. Make sure the data read out is in the same order as the data written into the FIFO. After reading the 16 "bytes," the control should change back to WRITE.

Questions:

1. How does the circuit provide safeguards against processor A inputting data to the FIFO while processor B is outputting data from the FIFO?
2. What condition is FIFO memory overflow? What condition is FIFO memory underflow? What problems are caused by the overflow and underflow?
3. What design and implementation considerations must be made in this organization for an 8-bit (real-byte) circuit? (Generally, 2 bits are called a dibit, 4 bits are called a nibble, and 8 bits are called a byte or octal.)

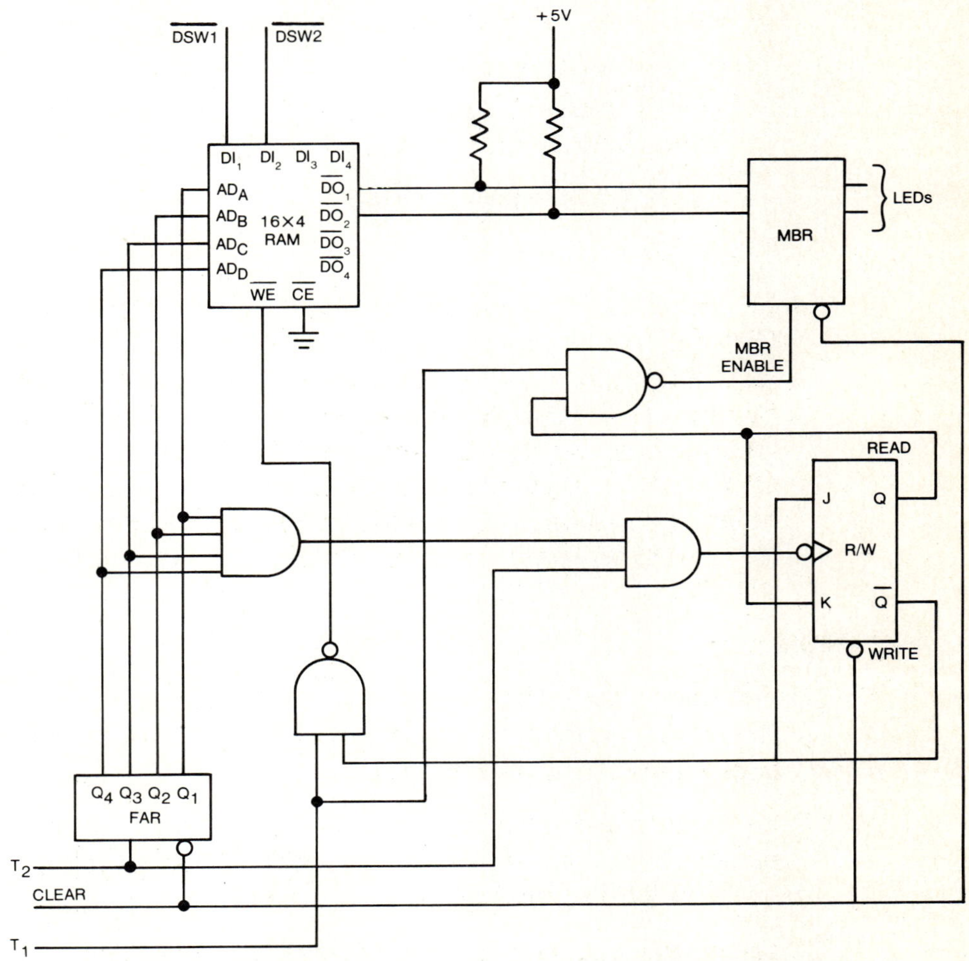

Figure 5-8. *The Circuit Diagram of the Hardware Communication Queue*

5.5 Experiment 11 - Communication Protocol Machine

Introduction:

A communication protocol is a set of mutually agreed upon conventions to regulate the information exchanges between communicating entities. The entities may be logic circuits, processors, terminals, software processes, or people. The communication path between communicating entities may be subject to noise, which corrupts the information being transferred. One of the most fundamental elements of a protocol is, therefore, to perform error-checking and recovery during the information exchange.

In this experiment, we introduce an error-detection technique that uses a redundant error-checking character at the end of every 15 data characters. Each error-checking bit of the error-checking character checks the parity of all the 15 corresponding bits of the data characters. This experiment expands on Experiment 10 to include the generation and checking circuits for the added error-checking character. There are also other types of error-detection techniques in communication protocols, such as the popular Cyclic Redundancy Check (CRC) code used in some commercial communication products.

Objectives:

1. to expand your knowledge of communication circuits, and
2. to develop an understanding of error-detection protocol requirements in a multiprocessor configuration.

Materials:

1. 7400 quad 2-input NAND gates.
2. 7420 dual 4-input NAND gates.
3. 7476 dual JK FF chips.
4. One 7489 16x4 RAM memory chip.
5. 10K-ohm resistors.

Equipment:

1. A logic trainer.

2. An extended breadboard. (Be sure to ground the two boards together.)

Procedures:

1. Design and implement a communication memory circuit with the following specifications. (Refer to Figure 5-9.)

   a. Basic Components:

      – a 16x2 FIFO memory, M,

      – a 4-bit FAR,

      – an MBR register,

      – a READ/WRITE indicator (a single FF is used to implement these two separate control signals),

      – a Longitudinal Redundancy Check (LRC) Character Generator (LRCCG), which is a 2-bit character implemented by two FFs,

      – an LRC Character Checker (LRCCC), which is a 2-bit character implemented by two FFs,

      – an LRC_GOOD indicator, which can be either a JK FF or a latch, and

      – an error generator control that can corrupt the data input to memory.

   b. FIFO Operations:

      – There are three operations associated with the FIFO: WRITE, READ, and CLEAR. Initially, the circuit is CLEARed. These operations are defined as follows:

WRITE

$WRITE$: $LRCCC \leftarrow 0$
$WRITE \cdot T_1$: if ($FAR \neq 15$) then
$M \leftarrow$ (data switches),
$LRCCG \leftarrow LRCCG \oplus$ (data switches)
else $M \leftarrow LRCCG$
$WRITE \cdot T_2$: $FAR \leftarrow FAR + 1$, $LRC\_GOOD \leftarrow 0$
if ($FAR = 15$) then
($WRITE \leftarrow 0$, $READ \leftarrow 1$)

READ

$READ$: $LRCCG \leftarrow 0$
$READ \cdot T_1$: $MBR \leftarrow M, LRCCC \leftarrow LRCCC \oplus M$
$READ \cdot T_2$: $FAR \leftarrow FAR + 1$
if ($FAR = 15$) then
($READ \leftarrow 0$, $WRITE \leftarrow 1$)
if ($FAR = 15$ AND $LRCCC = 0$)
then $LRC\_GOOD \leftarrow 1$

CLEAR

$CLEAR$: $FAR \leftarrow 0$, $WRITE \leftarrow 1$, $READ \leftarrow 0$,
$MBR \leftarrow 0$, $LRCCG \leftarrow 0$, $LRCCC \leftarrow 0$,
$LRC\_GOOD \leftarrow 0$

The WRITE operation is used to input data into the FIFO. The protocol requires that 15 data characters and one LRC character be transferred at a time. The LRC character is the exclusive-OR result of the input data being stored in the memory. All 16 bytes must be written into the FIFO before data can be removed. A READ operation is used to retrieve the data from the FIFO. Again, the protocol requires that all 16 bytes be read before new data can be written. The 16th byte is the check character and must be tested during the read to see if it results in a 0 in LRCCC. If it results in a 0 in LRCCC, then the LRC_GOOD indicator is turned ON to indicate that no error is detected for the transfer.

The CLEAR operation is used to initialize or reset the FIFO control to the WRITE mode and the FIFO address register to 0. Note that both WRITE and READ require timing signals T_1 and T_2 to control the sequence of operations. Additionally, from the micro-operations, we can see that T_2 can always increment FAR. The circuit diagram is suggested in Figure 5-10.

c. Error Operations:

– With no circuit faults or errors, the circuit will generate a correct LRC character and will not detect an error. The error generator can be built to alter the data stored in the memory. The altered data causes the LRC character check circuit to detect the error.

WRITE.error.T_1: If (FAR $\neq$ 15) then M $\leftarrow$ 00

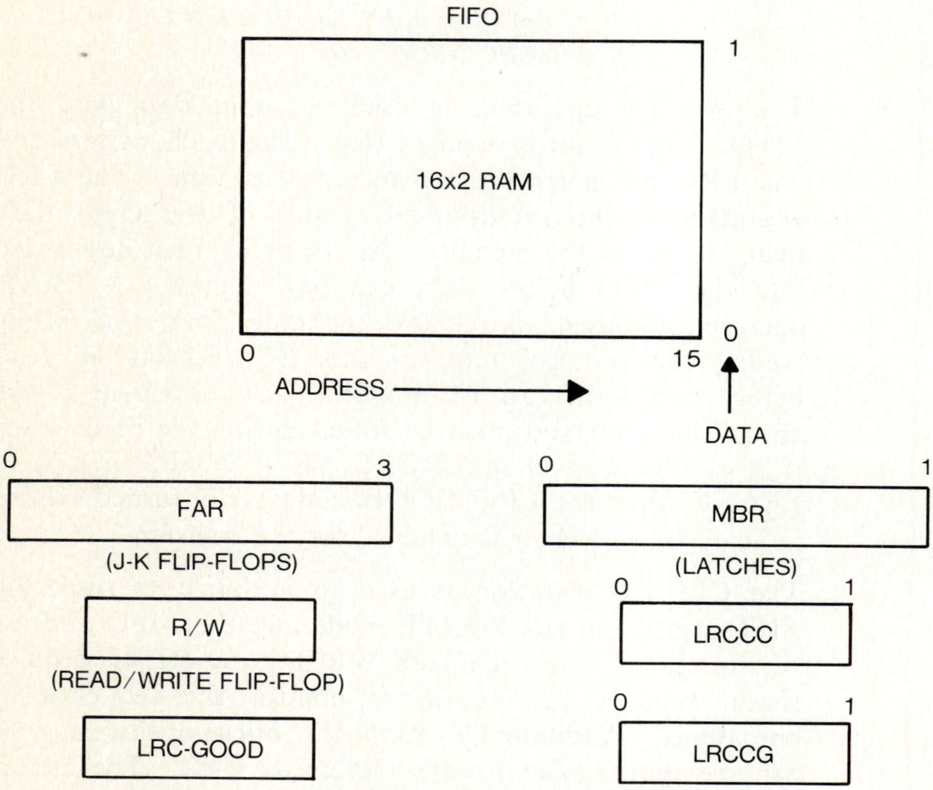

Figure 5-9. *The Basic Component Diagram of the Communication Protocol Machine*

2. Implementation Notes:

   – The 7489 RAM memory output requires 10K-ohm pull-up resistors.

   – Data into the memory from the data switches is inverted data as we have learned in the RAM experiment. Inverters can be used at the memory outputs to convert the data to the original values.

   – The control portion of the machine can be simulated by the pulser and data switches in order to generate the T_1, T_2, and CLEAR signals.

   – FAR is an UP counter.

   – READ and WRITE are two separate control signals and will never be a "1" at the same time. They are implemented by a single FF.

3. Testing Procedures:

   a. In order to demonstrate the FIFO operation, first CLEAR the circuit. The next step is to WRITE the values 00,01,10,11...00,01,10,11 until the FIFO is filled. On writing the 16th LRC character, the control should change the READ indicator to ON and the WRITE indicator to OFF. Make sure the data read out is in the same order as the data written into the FIFO. After reading the 16 characters, the control should change back to WRITE and NOT READ.

   b. If the LRC character being generated is the correct character, on the 16th read, the LRC_GOOD indicator should be ON. At all other times it should be OFF.

   c. Repeat Step a, but set the ERROR switch to LOW for one input that is either 01, 10, or 11. This will create the corruption of the data and force an error to occur. The LRC_GOOD indicator should remain OFF after the read to indicate the error.

Questions:

1. What is the function of the LRC character in the communication protocol?
2. Could the error be undetected if there is only one character of the 16 characters being corrupted?

3. What problems are caused by an LRC character that is in error? What problems are caused if there are errors in two data characters? Could the error be undetected by the circuit under these situations?

4. What design and implementation considerations must be made in this organization for an 8-bit (real-byte) circuit?

5. Modify the design of the circuit so that the content of the FIFO memory can be automatically retransmitted after an error is detected.

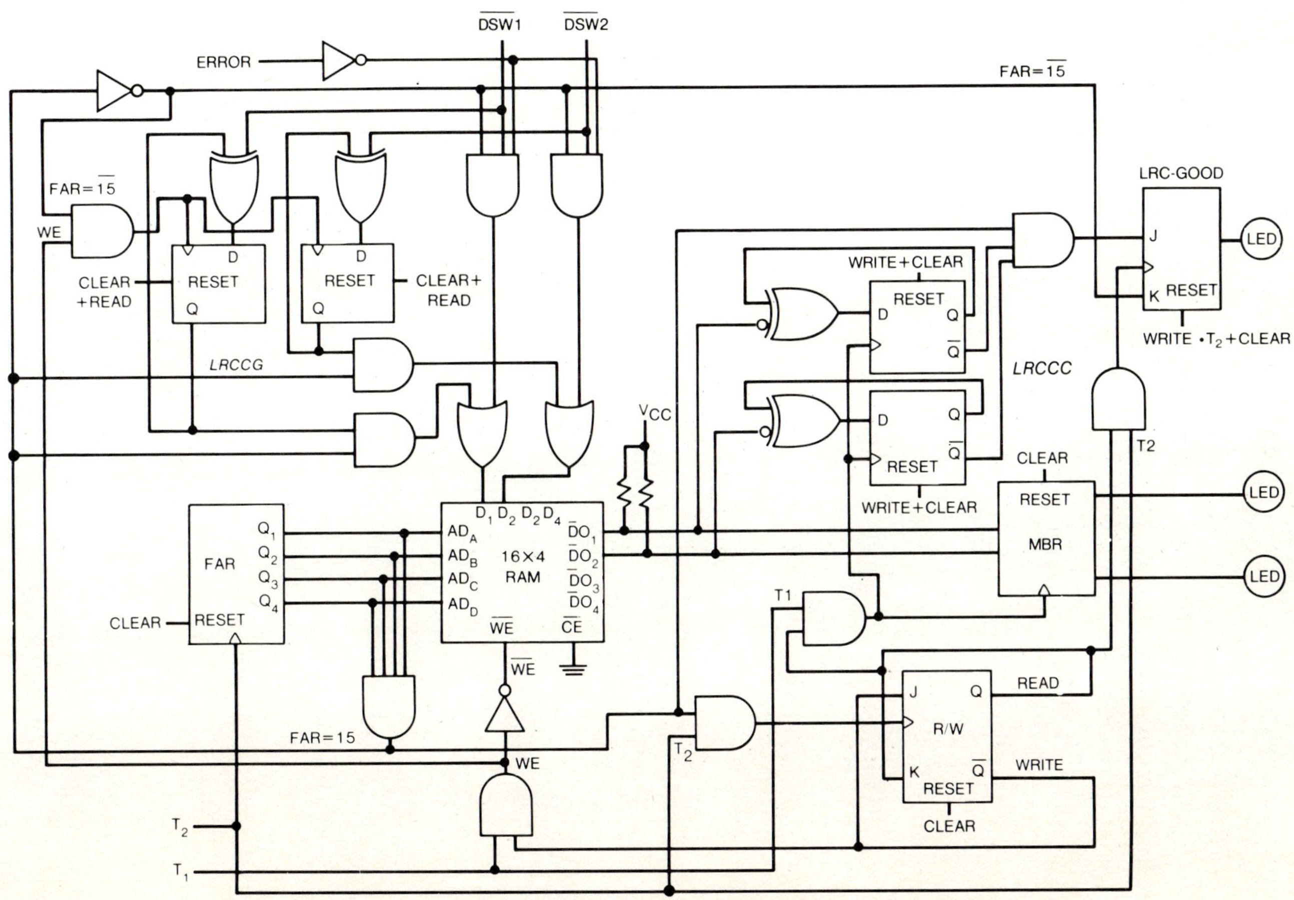

Figure 5-10. *The Basic Component Diagram of the Communication Protocol Machine*

5.6 Experiment 12 - Programmed Digital Controller

Introduction:

There are many types of programmed digital controllers available in the market. One of the applications is the microprogrammed control of the internal computer circuitry.

In this experiment, we build a simple controller that performs tasks stored in its memory. When completed, the controller will be free-running and will be tested with three separate programs. You can write other programs to simulate different types of operations.

Objectives:

1. to build a controller that can store and execute simple programs, and
2. to develop an understanding of the systematic steps needed to implement sophisticated circuits from the RT language description.

Materials:

1. 7400 quad 2-input NAND gates.
2. 7420 dual 4-input NAND gates.
3. 7475 4-bit bistable latches.
4. 7476 dual JK FF chips.
5. 7489 16x4 RAM memory chips.
6. 74154 4x16 decoder chips.
7. 10K-ohm resistors.

Equipment:

1. A logic trainer.
2. An extended breadboard. (Be sure to ground the two boards together.)

Procedures:

1. The basic organization of the system is shown in Figure 5-11. A layout is suggested in Figure 5-12.

2. This simple computer has two timing signals: T_1 and T_2.

3. The machine works in the following way:

$MBR \leftarrow M, PC \leftarrow PC + 1$
if $(MBR=0)$ then $AC_1 \leftarrow 1$
if $(MBR=1)$ then $AC_2 \leftarrow 1$
if $(MBR=2)$ then $AC_3 \leftarrow 1$
if $(MBR=3)$ then $AC_4 \leftarrow 1$
if $(MBR=4)$ then $AC_1 \leftarrow 0$
if $(MBR=5)$ then $AC_2 \leftarrow 0$
if $(MBR=6)$ then $AC_3 \leftarrow 0$
if $(MBR=7)$ then $AC_4 \leftarrow 0$
if $(MBR=8)$ then $AC \leftarrow 0000$
if $(MBR=9)$ then $AC \leftarrow 1111$
if $(MBR=10)$ then $PC \leftarrow 0$
if $(MBR=11)$ then $PC \leftarrow 0, AC \leftarrow 0000$
if $(MBR=12)$ then $PC \leftarrow 0, AC \leftarrow 1111$

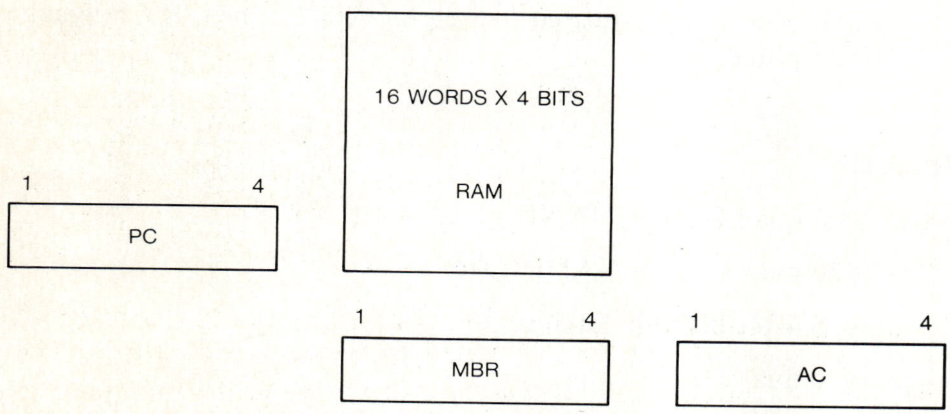

Figure 5-11. *The Basic Component Diagram of the Programmed Digital Controller*

4. Notice that MAR and Program Counter (PC) are assumed to be the same in this controller. A decoder can be used to decode the 4-bit instruction into separate signals Q_0–Q_{12}. PC is initially 0 and the Accumulator (AC) is connected to LEDs. The RT language description of the circuit is as follows:

T_1: $MBR \leftarrow M, PC \leftarrow PC + 1$
T_2Q_0: $AC_1 \leftarrow 1$
T_2Q_1: $AC_2 \leftarrow 1$
T_2Q_2: $AC_3 \leftarrow 1$
T_2Q_3: $AC_4 \leftarrow 1$
T_2Q_4: $AC_1 \leftarrow 0$
T_2Q_5: $AC_2 \leftarrow 0$
T_2Q_6: $AC_2 \leftarrow 0$

T_2Q_7: $AC_4 \leftarrow 0$
T_2Q_8: $AC \leftarrow 0000$
T_2Q_9: $AC \leftarrow 1111$
T_2Q_{10}: $PC \leftarrow 0000$
T_2Q_{11}: $PC \leftarrow 0000$, $AC \leftarrow 0000$
T_2Q_{12}: $PC \leftarrow 0000$, $AC \leftarrow 1111$

From the RT language description, a table of micro-operations by register can be derived as shown in Table 5-1. This table is helpful in designing the pieces of the circuits involving each register.

Table 5-1. *Micro-Operations by Registers*

Register	*Micro-Operation*	*When Performed*
MBR	$MBR \leftarrow M(LOAD)$	T_1
PC	$PC \leftarrow PC + 1$(Up Count)	T_1
"	$PC \leftarrow 0000$(Clear)	$T_2Q_{10} + T_2Q_{11} + T_2Q_{12}$
AC_1	$AC_1 \leftarrow 1$	$T_2Q_0 + T_2Q_9 + T_2Q_{12}$
"	$AC_1 \leftarrow 0$	$T_2Q_4 + T_2Q_8 + T_2Q_{11}$
AC_2	$AC_2 \leftarrow 1$	$T_2Q_1 + T_2Q_9 + T_2Q_{12}$
"	$AC_2 \leftarrow 0$	$T_2Q_5 + T_2Q_8 + T_2Q_{11}$
AC_3	$AC_3 \leftarrow 1$	$T_2Q_2 + T_2Q_9 + T_2Q_{12}$
"	$AC_3 \leftarrow 0$	$T_2Q_6 + T_2Q_8 + T_2Q_{11}$
AC_4	$AC_4 \leftarrow 1$	$T_2Q_3 + T_2Q_9 + T_2Q_{12}$
"	$AC_4 \leftarrow 0$	$T_2Q_7 + T_2Q_8 + T_2Q_{11}$

The one remaining problem to solve is how to write a program into memory with the minimal amount of effort. The micro-operations needed to accomplish this are as follows:

T_1: $MBR \leftarrow M$, $PC \leftarrow PC + 1$
T_2: $M \leftarrow$ Input Data Switches

The T_1 operation described above has not changed. The T_2 change is easily accomplished by disconnecting logic switch 2 to the WE input of the memory chip. Writing the program to memory requires that the clock be connected to the logic switch 1 so that the input switches can be reset before each T_2 pulse.

5. Write a program to display the following counting sequences on the AC displays:

Sequence 1:

$0 \rightarrow 1 \rightarrow 0 \rightarrow 2 \rightarrow 0 \rightarrow 4 \rightarrow 0 \rightarrow 8 \rightarrow 0 \rightarrow 1$ ...

Sequence 2:

0 → 1 → 9 → 11 → 15 → 14 → 6 → 4 → 0 → 1 ...

Sequence 3:

0 → 15 → 7 → 3 → 1 → 5 → 15 → 0 → 15 ...

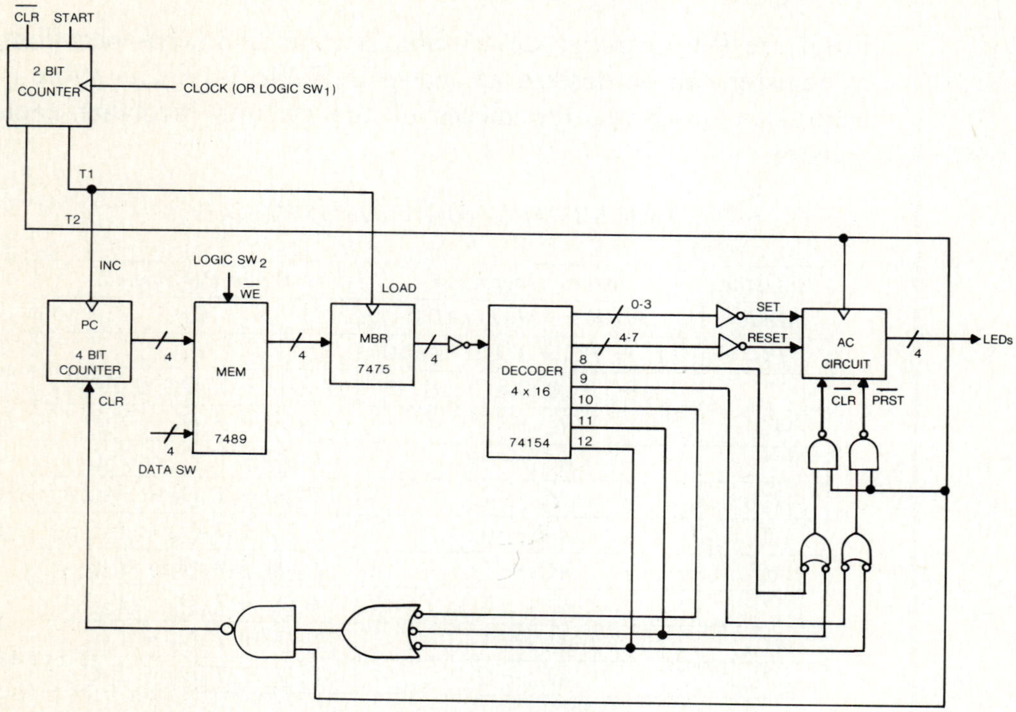

Figure 5-12. *The Circuit Layout of the Programmed Digital Controller*

5.7 Design Exercise 1 - An Odd-Even Counter

Design and implement a 2-bit counter that counts as follows:

- The counter has two control inputs called P_1 and P_2.
- When $P_1 = 0$, the counter counts in the sequence 0,1,3; ....; 0,1,3; ....
- When $P_1 = 1$, the counter counts in the sequence 0,2; ....; 0,2; ....
- When $P_2 = 0$, the state of the counter does not change.
- When $P_2 = 1$, the counter counts in the sequence determined by P_1.

Note: If the counter is counting according to the $P_1 = 0$ sequence and P_1 changes to 1, the counter should reset to 0 and count recording to the $P_1 = 1$ sequence. The counter behaves similarly if P_1 changes from 1 to 0.

5.8 Design Exercise 2 - A Furnace Temperature Controller

Design a hypothetical furnace controller that can perform the following tasks:

1. The controller accepts a Desired Room Temperature (DRT) inputted by enabling the Set_temp switch.
2. The Room Temperature (RT) may range from 0-15°F and the DRT may be set between 0-13°F. The controller's task is to efficiently turn on and turn off the heating unit and the circulation fan unit to heat the room to the DRT.
3. Every clock time (1 second) the controller gets the current room temperature from its sensor. It then enables the RT and DRT comparison at every clock time.
4. If the room is not warm enough and the fan is off, the heating unit will be turned on. The fan will be turned on 4 seconds after the heating unit is turned on to allow for a warm-up period.
5. If the room is already warm enough and the fan is on, the heating unit will be turned off. The fan will be turned off 4 seconds after the heating unit is off to transfer more heat.

    a. Design the controller using a minimum number of components. Draw the circuit diagram in a block diagram form. (Assume that a 4-bit comparator is available that compares two 4-bit numbers, X and Y, and determines their relative magnitudes. The outcome of the comparison is specified by three outputs that indicate if $X > Y$, $X = Y$, or $X < Y$.)

    b. Describe the furnace controller design using RT language.

    c. Write down the input/output connections of the furnace controller to the logic trainer for implementing your design.

    d. Write the major steps for testing your implementation.

5.9 Design Exercise 3 - A Multiplication Circuit

Design a sequential circuit that performs the 2-bit multiplication. The multiplication circuit should be implemented by a series of shifts and additions. It requires a 4-bit storage to store the result.

1. Describe your design using RT language.
2. Implement the circuit utilizing NAND gates, JK FFs, and/or registers.

3. Draw the circuit diagram including connections to the logic trainer.

4. List the steps for testing your implementation.

5. Describe a design that can perform n-bit multiplication using RT language.

5.10 Design Exercise 4 - An Asynchronous Communication Circuit

An asynchronous communication circuit can transmit data asynchronously. It allows the transmission of each byte to be individually synchronized, usually by the use of start and stop bits. The idle time between each byte is not necessarily fixed clock periods. The majority of the low-speed systems, such as computer terminals, use asynchronous communication circuits.

The transmit portion of the circuit loads a data byte into its transmit buffer register. It then inserts a prefix bit (called start bit) to the data bits of the transmit buffer register to indicate the start of the byte and inserts one bit (called stop bit) as a suffix bit to the register to indicate the end of the data byte. At the receive portion of the circuit, the start bit can activate the receive circuit to receive the following data byte and the stop bit can deactivate the receive circuit to indicate data received. The transmit and receive circuits are controlled by different clocks of the same frequency.

A Universal Asynchronous Receiver/Transmitter (UART) is an MSI device commonly used in data communication terminals, controllers, and computer interface circuits. A Harris HD-6402 chip UART is available to perform asynchronous communication (see Reference 5 of Logic Design in the Bibliography). Design a simple communication circuit that meets the following specifications:

Transmit Circuit

- Inputs
  - Two Data Inputs (DI_1, DI_2).
  - One Transmit Buffer Register Load (TBRL).
  - One Transmit Register Clock (TRC).
  - One Reset.
- Outputs
  - One Serial Output Data (SOD).
  - One Transmit Buffer Register Empty (TBRE).

- Operation
  - A HIGH applied to "Reset" causes SOD and TBRE to go HIGH and stops transmitting.
  - If TBRE and SOD are HIGH, output data byte can be applied to DI_1 and DI_2. After applying data to DI_1 and DI_2, clock data into the Transmit Buffer Register using TBRL. The leading edge of the clock pulse gates the data into the transmit buffer register and the trailing edge of the clock pulse sets TBRE to a LOW and starts the transmit cycle.
  - The transmit cycle causes a LOW output (start bit) for the first output cycle. The second and third output cycles are data bits 1 and 2, respectively. The fourth output cycle causes a HIGH (normal idle output as stop bit) at SOD, sets TBRE HIGH, and ends the transmit cycle. (See Figure 5-13.)
  - The TRC transmitter clock *must be* the same frequency as the receiver's RSC clock. Normally, the clocks are separate, independent signal sources and are located "within" the respective transmitter/receiver device. The implementation of this exercise can use the same signal source.

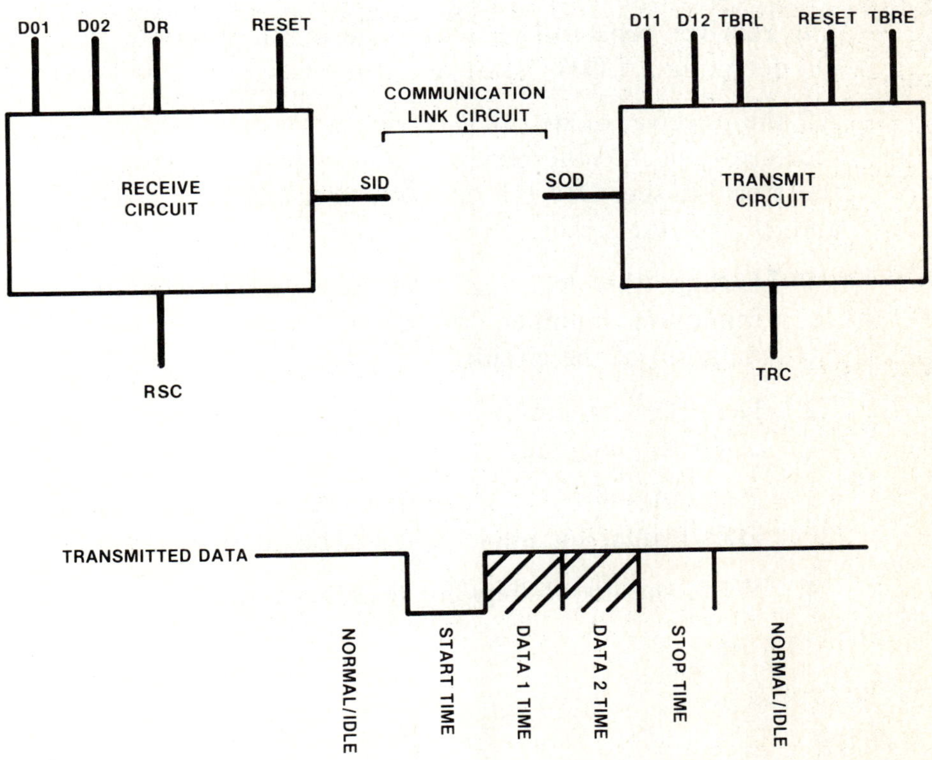

Figure 5-13. *An Asynchronous Receiver/Transmitter Circuit*

Communication Link Circuit

- Connect the transmit SOD to receive SID.
- If the transmit and receive circuits are on separate breadboards, connect a ground from the transmit board to a ground on the receive board.

Receive Circuit

- Input Leads
  - One Serial Input Data (SID).
  - One Receive Serial Clock (RSC).
  - One Reset.
- Output Leads
  - One Data Received (DR).
  - Two Data Out Leads (DO_1, DO_2).
- Operation
  - A HIGH applied to the reset lead sets DR LOW, sets idle mode (not receive mode), and DO_1 and DO_2 to a LOW.
  - The receiver constantly samples the input (SID) for a LOW. On detecting a LOW (start bit), the receiver sets DR LOW.
  - At the receive mode, the receive circuit serially shifts the received data into the receive buffer register. The reception of the stop bit sets DR HIGH (DO_1 and DO_2 reflect data bits 1 and 2) and resets receive mode. (See Figure 5-13.)

1. Design the transmit and receive circuits using the minimum number of components. Draw a block diagram of the circuit.
2. Describe separately the transmit and receive circuits using RT language.
3. Draw the circuit in detail indicating connection on the logic trainer for implementing your design.
4. Write the major steps for testing your implementation.

6

Digital System Projects

From the instructional and design experiments, we have learned many important building blocks that can be used to construct challenging digital system projects. We cover three design projects in this chapter for sophisticated LSI/VLSI logic design: a microcomputer, a VLSI sorting engine, and a computer network interface. Due to the complexity of implementing these projects, you may choose simply to study the principles of the projects without attempting to implement them.

From the following three design projects, you can gain some insights about the internal organization of a digital computer, the trend of hardware implementation of software functions, and the principles of computer networks and their protocols.

The projects illustrate three important areas for hardware applications. Each design is built by a composite of the building blocks introduced in the previous experiments. You should be able to explore other areas of IC applications at the end of the study.

The projects cover a wide range of topics, and we recommend you supplement the projects with lectures or by reading material/lectures from References 2, 3, 4, 5, and 7 of Computer Design in the Bibliography.

6.1 Project 1 - A Simple Microcomputer Design

This project implements a hypothetical computer, model 470/01. Before we describe the design for it, we will first present an overview of microcomputers and their applications.

6.1.1 Microcomputer Overview

A microcomputer is a complete computer system that uses a microprocessor as its CPU and includes memories, peripheral input/output devices, and a power supply. There are a large number of applications for microcomputers such as traffic light controllers, toys and games, robots, electronic appliances, and general-purpose computers.

To understand the organization of a microcomputer system, let us consider the following areas of studies: 1) word size for each address unit, 2) memory type and configuration, 3) register, 4) instruction format, 5) addressing mode, 6) instruction set, 7) control unit, and 8) input/output devices of the computer. The following is a brief discussion of the topic related to this project. If you need more information, you can study the microcomputer in detail from one of the computer architecture textbooks.

After you understand the microcomputer organization, we will actually design the internals of a simple hypothetical computer, model 470/01, in this project. It is a computer specifically intended for studying the fundamentals of computer design.

6.1.1.1 Word Size

A word in computer memories is the basic address unit. It decides the size of storage unit for data and instructions of computer programs.

There are many different sized microprocessor chips available in the market such as an 8-bit word microprocessor (Intel 8080, Motorola 6800, etc.) and a 16-bit word microprocessor (Intel 8086, Zilog Z8000, etc.). There is also a 32-bit microprocessor today that has more computing power than the most expensive computer in the early 1950s. The instruction word size limits the directly addressable memory size. A large size microprocessor can accommodate large programs more efficiently than a small microprocessor.

6.1.1.2 Memory

We have examined the memory topic in Experiment 6 with the introduction of RAM memory circuits. The content of RAM is lost when the power is turned off. ROM, on the other hand, is not volatile and is usually preprogrammed by the manufacturer. The memory size of a computer is a key factor in rating its flexibility and power.

6.1.1.3 Registers

There are many types of registers such as PC, Instruction Register (IR), and AC register. The PC keeps the address of the next instruction to be executed in a computer. The IR may be the same as an MBR and is used to store the current instruction during its execution. An AC is a common register used for arithmetic and logic manipulation of data. Most computers have a special set of general registers (such as stack, base, and index registers) to 1) speed data access and processing or 2) increase address space and decrease instruction length.

6.1.1.4 Instruction Format

Generally, an instruction consists of Operation (OP) code and instruction Addresses (ADR). The operation is to be performed on operand(s) specified by the ADRs. In some cases, it is implied that one or more of the operand addresses is the accumulator. For example, the binary Addition (AD) operation of the 470/01 computer requires the operand specified by ADR to be added to AC, and the result then put into AC.

6.1.1.5 Addressing Mode

The address used to fetch the actual data is generally referred to as the EFFECTIVE ADDRESS (EFFADR). In direct addressing mode, the address part of the instruction (ADR) is equal to EFFADR. In indirect addressing, ADDR is the address of a data that is the effective address. That is EFFADR=M(ADR). Thus, ADR references a pointer that gives the location of the actual operand to be processed. An indirect bit is needed in the instruction format to show the indirect addressing. Some machines allow multiple levels of indirections. Indirection is very useful for subroutine linkage and to enhance addressing space.

There are many other addressing modes with different numbers of operands for instructions. Question 2 of this project asks you to compare different addressing modes and their impact on program sizes.

6.1.1.6 Instruction Set

The specified instruction to be performed is selected by the OP code portion of the instruction. There are three basic types of instruction: transfer, ALU, and control.

Transfer instructions move data among memory words and registers without changing the content. Examples are LOAD, STORE, EXCHANGE, PUSH, POP, INPUT, and OUTPUT.

ALU instructions perform arithmetic and logic operations on data stored in registers or memory words. Examples are arithmetic operations (such as addition, subtraction, multiplication, and division), logic operations (such as AND, OR, NOT, and XOR), and shift operations (such as right shift, left shift, and rotate shift).

Control instructions test conditions in registers, and, depending on results, cause a change in program execution sequence. Examples are JUMP (BRANCH), CALL/RETURN, and SKIP. The change of execution sequence can be done unconditionally or only when some required conditions are met. The PCs are usually updated to fetch the next instruction so that instructions are executed sequentially. A control-type instruction can change the sequence of execution. For example, a branch instruction is accomplished by providing the desired address in EFFADR of the instruction. The EFFADR is moved to PC and causes the next instruction to be taken from the desired address.

6.1.1.7 Control Unit

Most commercially available microprocessors work in the following way:

1. First of all, when the power is turned on, the microprocessor's PC is initialized to address the starting memory location of a program.
2. The microprocessor reads the content of the addressed word from the memory bus and puts the word into its internal register(s) such as MBR, AC, or IR. The microprocessor also reads in the address for the operand if the instruction requires indirect addressing.
3. The instruction is decoded in the control unit of the microprocessor so that the microprocessor knows whether it is an addition, shift, load, store, or branch instruction.
4. The decoded instruction activates the corresponding control pulses in the control unit to execute the instruction by using the arithmetic and logic unit of the microprocessor.
5. The control unit updates the program counter to point to the next instruction to be executed.
6. The process is repeated between Step 2 and Step 5.

Most instructions require a sequence of micro-operations to be performed. For example, the indirect addressing mode of the LOAD instruction has multiple memory fetch operations. We often use a hardware description language (RT language) to specify the sequence of events to execute the instruction. For optimal processor speed, many operations that can be executed in parallel are combined in one control clock. A control unit implements the hardware description of the design and enables the correct set of transfers and arithmetic or logic operations in the desired sequence. The execution speed of computer instructions is also a key factor in rating the power of a computer.

There are two approaches to implementing the control unit. The first approach is via microprogrammed control. The sequence of control is stored in control memory in the format of microinstructions. Microinstructions are executed to perform the machine's instruction. This first approach is flexible for modifying the design but is slow due to the fetch time required from control memory. The second approach is via hardwired control. Hardwired control is implemented directly in a large sequential circuit. The second approach is less flexible but faster than the first approach.

In this project, we use the RT language to describe a hardwired control unit.

6.1.1.8 Input/Output

The input/output devices connected to a microcomputer may be switches, joy sticks, keyboards, modems, terminals, tapes, or disk devices, etc. Switches may be data switches or pulser switches to send binary data to the computer. A joy stick is often used to play video games. A keyboard can be used to send character messages to the computer. A modem can link the computer to the outside world via voice telephone lines. A terminal can be either a paper-copied printer and/or a screen-displayed monitor. A tape device may be a regular cassette tape or a large computer magnetic tape, and a disk may be floppy or hard. Both tape and disk devices can be used to store large quantities of data and are usually called secondary storages while the random access memory is called the primary storage. More advanced devices include optical scanners, light pens, speech synthesizers, etc.

Generally, there are I/O interfaces between computers and I/O devices. When a CPU wants to move a block of data items from/to an I/O interface, it must wait until the I/O device obtains one data item before receiving/sending another data item. An I/O interface contains some status registers that can be checked by the CPU to determine the status of a previous I/O operation. There are three methods of getting the I/O status to perform an I/O operation: 1) program-controlled I/O, 2) interrupt-driven I/O, and 3) direct memory access. You may want to refer to these techniques in a standard computer architecture book.

The 470/01 computer handles input only from data switches, and sends its output to the accumulator.

A Motorola 6800-based microcomputer trainer and a series of experiments are available for experimentation using microprocessors. Experiments using commercially available microprocessors are beyond the scope of this book and we recommend that you investigate the subject on your own.

6.1.2 Microcomputer Applications

Hundreds of microcomputers are available in the market and the selection of a microcomputer, like selecting a car, depends on the price, the available applications, the speed and capacity, and individual taste. Microcomputer hardware is useless without the supporting software programs to expand its applications. The programs that are common to a microcomputer are operating systems, assemblers, compilers, simulators, utility routines such as text editors, and application programs such as accounting packages.

The software development for a microcomputer is often much more involved than the design of its hardware. Due to the limited capacity of a microcomputer, a sophisticated software development system is often built on a large mainframe computer for the microcomputer. The development system on the large development machine is easy to use and can simulate the microcomputer. The developed software is then downloaded to the microcomputer for the final testing and the actual application.

We will describe the design of a simple microcomputer for the rest of the project by using the principles outlined in the microcomputer overview.

6.1.3 Design Description

Objectives:

Design with pencil and paper only, a hypothetical 470/01 computer system. This machine is not intended to serve as a commercially available microcomputer. It is a highly simplified design for the purpose of teaching the design process of a complete computer system.

Materials:

1. 7400 quad 2-input NAND gates.
2. 7420 dual 4-input NAND gates.
3. 7475 4-bit bistable latches.
4. 7476 dual JK FF chips.
5. 7486 quad 2-input XOR gates.
6. One 7489 16x4 bit memory.
7. One 74154 4x16 decoder chip.
8. 10K-ohm resistors.

Procedures:

1. The organization of this computer system is shown in Figure 6-1.

2. Computer Architecture Description:

   We will describe the 470/01 computer architecture by using the same eight areas of studies as shown in the previous section.

   a. Word Size: 4 bits

   b. Memory Size: 16 words RAM

   c. Registers:

Mnemonic	*Register*	*Size*
AC	Accumulator Register	4 bits
IR	Instruction Register	3 bits
I	Indirect FF	1 bit
PC	Program Counter	4 bits
MAR	Memory Address Register	4 bits
S	Stop FF	1 bit

   d. Instruction Format:

0 1 2	3	4 5 6 7
OP	I	ADR

   e. Addressing Mode: Direct and Indirect Memory Addressing

   if (I=0) then EFFADR = ADR
   if (I=1) then EFFADR = M(ADR)

f. Instruction Set:

OP	*Mnemonic*	*Function*
0	HT	Halt
1	IN	Load AC with Input Data Switches
2	SH	if (I=0) then AC Right Shift
		if (I=1) then AC Left Shift
3	LD	Load AC with M(EFFADR)
4	ST	Store Content(AC) at EFFADR
5	AD	Add M(EFFADR) to Content(AC)
6	BU	Unconditionally Branch to EFFADR
7	BZ	if (AC=0) then Branch to EFFADR

The length of an instruction can be either one word or two words. If OP < 3 then only one word is used for each instruction. During the execution of an instruction, PC should be incremented by the length of the current instruction. This way, PC is always the address of the next instruction after each instruction execution.

g. Control Unit:

The 470/01 computer works as described in Section 6.1.1.7 and uses the hardwired implementation approach. Each instruction of the 470/01 computer requires a sequence of micro-operations to be performed and these micro-operations are executed in four clock cycles (C_0–C_3). The four cycles are called: 1) fetch the instruction, 2) fetch the operand address, 3) fetch the address for an operand in an indirect addressing mode, and 4) execute the instruction. The cycle (state) diagram for the 470/01 is shown in Figure 6-2 with two FFs, F and R, representing these four states.

h. Input/Output

The inputs of the microprocessor are data switches and pulser switches. The outputs of the microprocessor are LEDs.

The system circuit does not have a manual system for loading the memory with a test program. This is done to keep the circuit as simple as possible.

3. Design Steps

Figures 6-3 through 6-5 outline the major design steps and circuits for the 470/01 computer. The CPU has four cycles

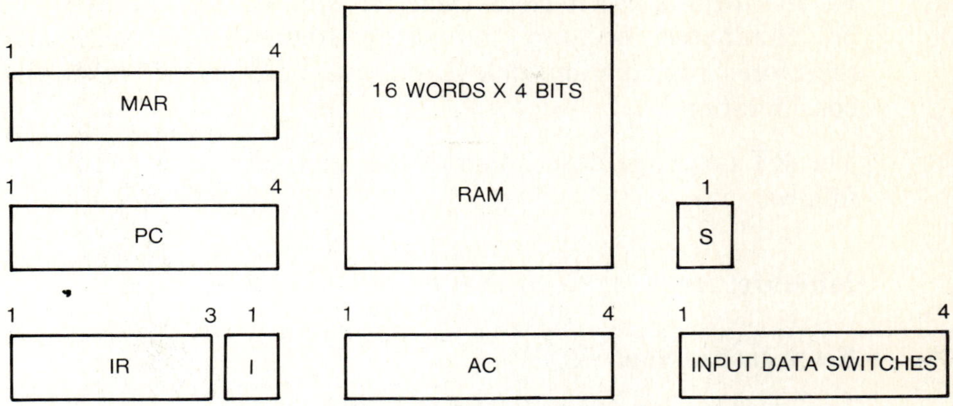

Figure 6-1. *The Basic Component Diagram of the Microcomputer*

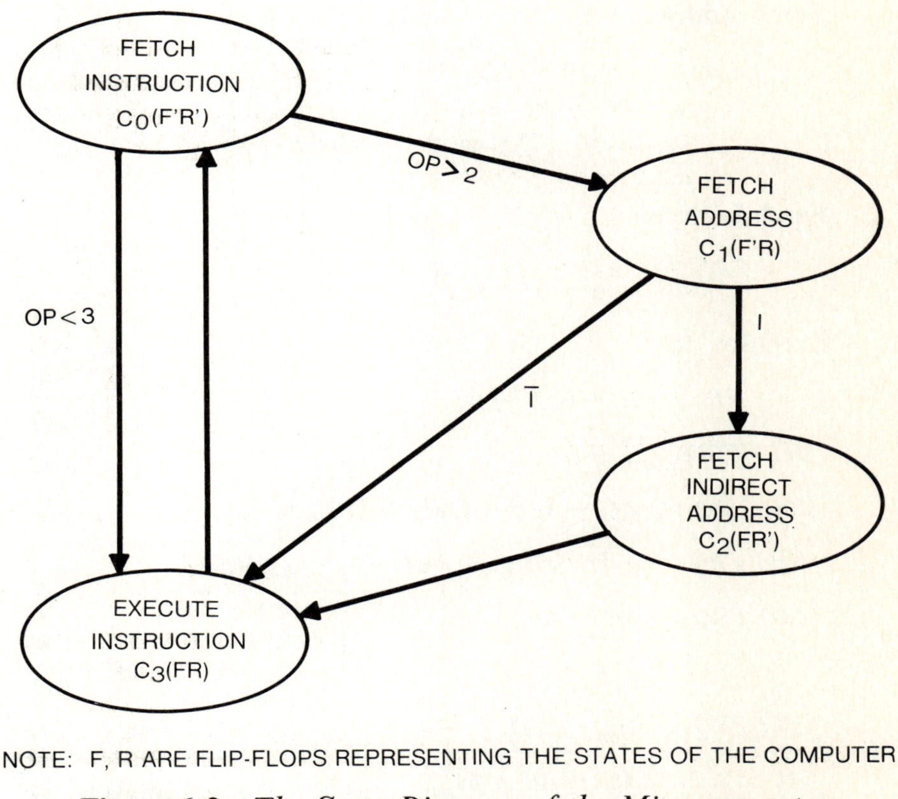

Figure 6-2. *The State Diagram of the Microcomputer*

and each cycle needs only one timing signal. Memory addressing can be selected either from PC or MAR. This design can eliminate the register transfer between MAR and PC and results in a 1-clock cycle for each state. The MBR can be eliminated because the clock frequency is very slow, therefore, propagation delays do not need to be taken into consideration.

The RT language description of the microcomputer circuit is as follows:

Initialize: $PC \leftarrow 0, F \leftarrow 0, R \leftarrow 0$

Fetch Instruction: (C_0)

$F'R'$: $IR \leftarrow M_{pc}(OP), I \leftarrow M_{pc}(I),$
$PC \leftarrow PC + 1, R \leftarrow 1,$
if $(OP > 2)$ then $F \leftarrow 0$ else $F \leftarrow 1$

Fetch Address: (C_1)

$F'R$: $MAR \leftarrow M_{pc}, F \leftarrow 1,$
$PC \leftarrow PC + 1,$
if $(I = 1)$ then $R \leftarrow 0$ else $R \leftarrow 1$

Fetch Indirect: (C_2)

FR': $MAR \leftarrow M_{mar}, R \leftarrow 1, F \leftarrow 1$

Execute: (C_3)

FR: $R \leftarrow 0, F \leftarrow 0$

$(HT)\ FRQ_0$: $S \leftarrow 0$

$(IN)\ FRQ_1$: $AC \leftarrow$ Input Data Switches

$(SH)\ FRQ_2$: if $(I = 0)$ then $SHR \leftarrow 1$ else $SHL \leftarrow 1$

$(LD)\ FRQ_3$: $AC \leftarrow M_{mar}$

$(ST)\ FRQ_4$: $M_{mar} \leftarrow AC$

$(AD)\ FRQ_5$: $AC \leftarrow AC + M_{mar}$

$(BU)\ FRQ_6$: $PC \leftarrow MAR$

$(BZ)\ FRQ_7$: if $(AC = 0000_2)$ then $PC \leftarrow MAR$

From the RT language description, a table of micro-operations by registers can be derived as shown in Table 6-1. This table is helpful in designing the various pieces of the circuits involving each register. The control unit circuit can also be derived from the table and is partially shown in Figure 6-5. The portion of the control unit circuit that is not completed is left as an exercise for you.

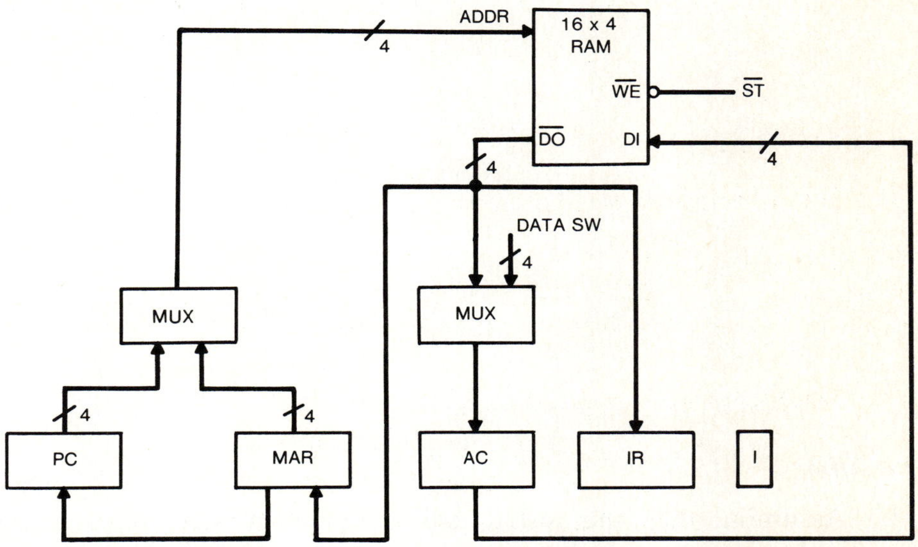

Figure 6-3. *The Block Diagram of the Microcomputer*

Table 6-1. *Micro-Operations by Registers*

Register	*Micro-Operation*	*When Performed*
PC	$PC \leftarrow PC + 1$	$C_0 + C_1$
PC	$PC \leftarrow MAR$	$C_3(Q_6 + Q_7(AC = 0000))$
IR	$IR \leftarrow M_{pc}(OP)$	C_0
I	$I \leftarrow M_{pc}(I)$	C_0
MAR	$MAR \leftarrow M_{pc}$	C_1
MAR	$MAR \leftarrow M_{mar}$	C_2
Memory	$M_{mar} \leftarrow AC$	$C_3Q_4(ST)$
AC	$AC \leftarrow$ Input Data Switches	$C_3Q_1(IN)$
AC	$AC \leftarrow SHR(AC)$	C_3Q_2I' (SH)
AC	$AC \leftarrow SHL(AC)$	C_3Q_2I
AC	$AC \leftarrow M_{mar}$	$C_3Q_3(LD)$
AC	$AC \leftarrow AC + M_{mar}$	$C_3Q_5(AD)$
S	$S \leftarrow 0$	$C_3Q_0(HT)$
F	$F \leftarrow 0$	$C_3 + C_0(OP > 2)$
F	$F \leftarrow 1$	$C_0(OP < 3) + C_1 + C_2$
R	$R \leftarrow 0$	$C_3 + C_1I$
R	$R \leftarrow 1$	$C_0 + C_1I' + C_2$

4. Testing Procedures:

Finish the rest of the design for the 470/01 computer and trace the circuit by using the following test program.

Memory Address	*Word*
0	0010
1	0101
2	0100
3	1010
4	0111
5	1000
6	1101
7	0110
8	1100
9	1110
10	1101
11	0010
12	0000
13	xxxx
14	1111
15	1011

Questions:

1. Assuming that the restrictions on memory size, register size, and input switches are removed from the 470/01 implementation, write a program that can calculate N!, where N = 1 to 128. What is the register size required? How many input switches are required?

2. Given the following four CPU organizations, determine the time each CPU organization will need to execute a program to calculate the following:

   X = A/(B*C-D*E) where A, B, C, D, E, and X are in memory and an instruction (such as multiplication, division, subtraction, etc.) takes 1 μs for each organization.

   CPU A: Three memory addresses per instruction, CPU operations on two operands, and store the result in the third.

   CPU B: Two memory addresses per instruction, CPU operations on the two operands, and store the result in one of them.

   CPU C: Single accumulator, one memory address per instruction, and store the result in the accumulator.

CPU D: Stack machine, CPU operations to the top two registers in the stack, and PUSH/POP operations on the stack.

3. A very simple "computer" is available without a PC. Instead, all instructions contain four parts: an operation code (for addition and subtraction operations), two accumulators (A and B), an address of an operand, and the address of the next instruction. The operations are performed on one of the accumulators and the operand, and the result is stored in the accumulator.

   a. Draw a block diagram and the logic circuit diagram of the "computer."

   b. Write a sequence of micro-operations (with control functions) for both fetch and execute cycles of the "computer."

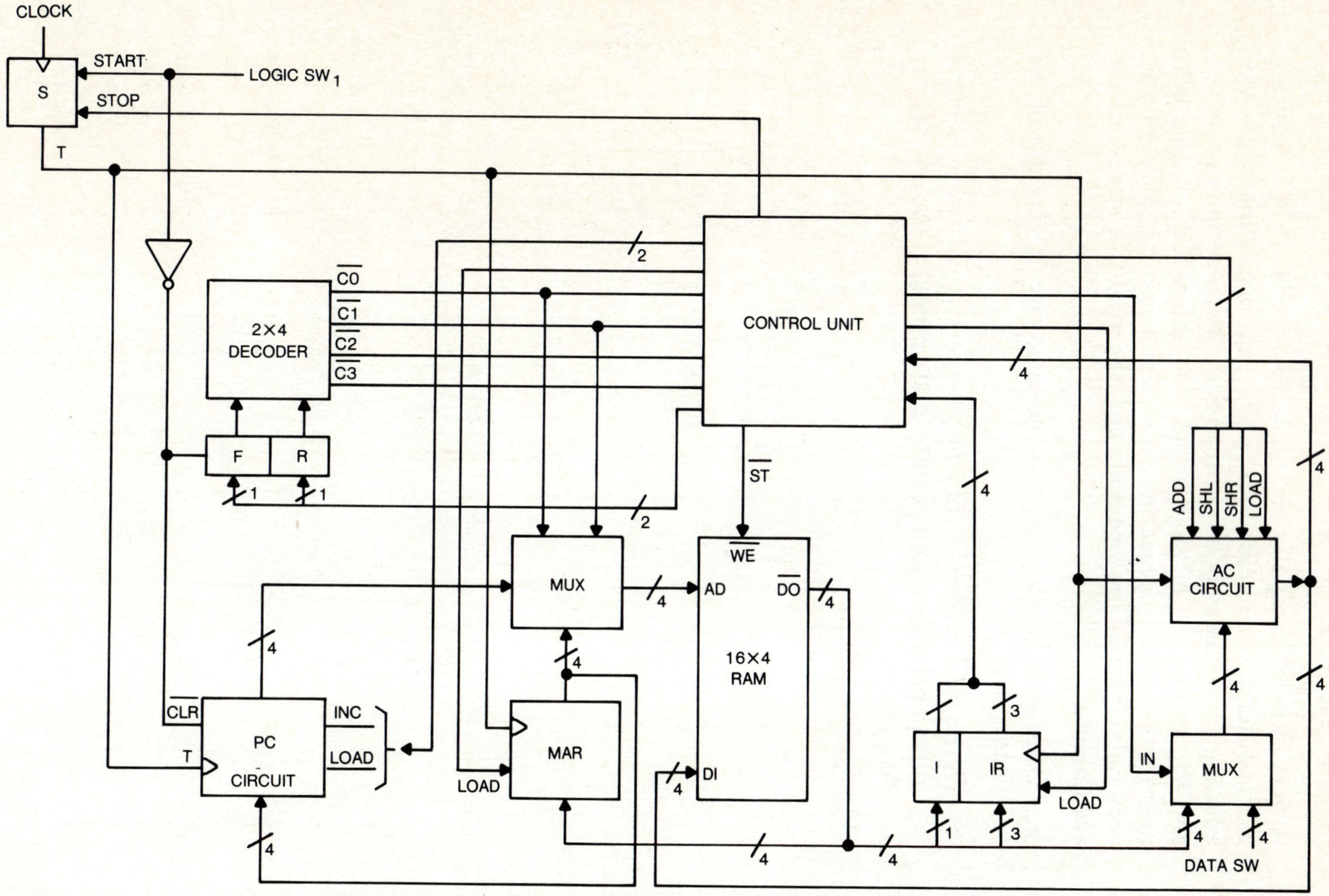

Figure 6-4. *The Basic Circuit Layout of the Microcomputer*

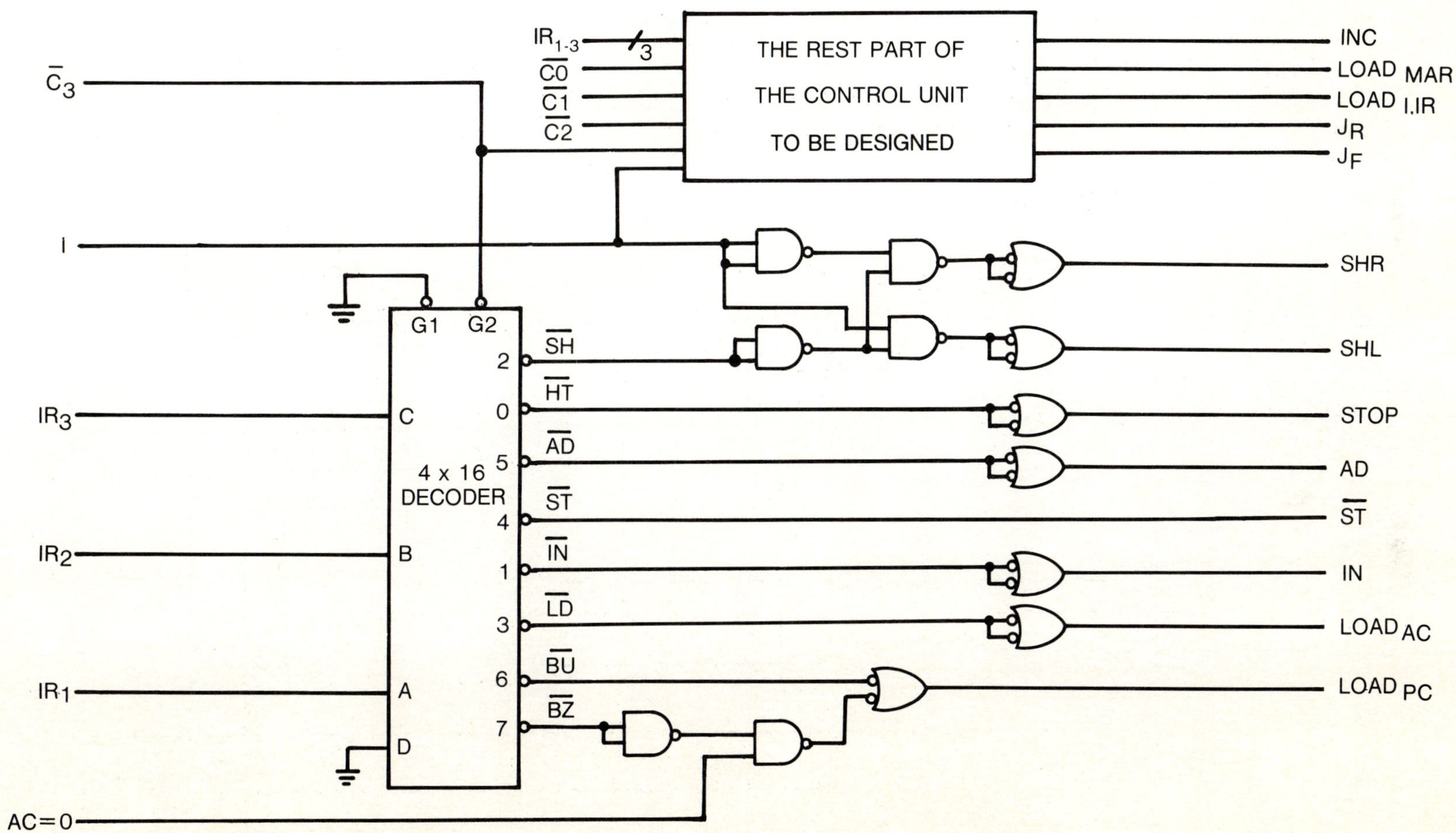

Figure 6-5. *The Control Unit of the Microcomputer*

6.2 Project 2 - A VLSI Sorting Engine

6.2.1 VLSI Overview

This project shows a VLSI sorting engine that can be used efficiently to perform sorting by hardware. The design is adopted from a paper, "An ON-CHIP Compare/Steer Bubble Sorter," by D. T. Lee, H. Chang, and C. K. Wong, published in *IEEE Transactions on Computers*, June, 1981.

As a result of the fierce competition in fabrication technology by leading IC manufacturers, VLSI circuitry has become so dense that a single silicon VLSI chip may contain millions of transistors. VLSI electronics present a challenge, not only to those involved in the fabrication technology, but also to computer scientists and computer architects.

VLSI designers usually have to use prototype chips and go through many trials before all the circuit design problems are resolved. The sophisticated VLSI design demands the use of software techniques to aid the design and to eliminate many of the prototype-debugging processes.

A sorting engine can be implemented by VLSI circuitry. It consists of a large number of storage units with the associated sorting logic. A processing unit can start the sorting engine and use it for sorting in a very simple and efficient way. The processing unit can simply throw a sequence of unsorted numbers into the engine and then immediately take a sequence of sorted numbers out of the engine. The sorting is completed within the input plus output time. The speed is, therefore, extremely fast and is only limited by the input and output clock frequencies. With this kind of engine, the time-consuming sort operations by software can be completely eliminated.

With the economy of VLSI circuitry, many other software operations may be implemented by hardware in the future. This project demonstrates the hardware circuitry for realizing the VLSI sorting engine. If you are in a laboratory course you are encouraged to implement a unit of the circuitry and then connect it with other implemented units as an experiment.

6.2.2 Design Description

To design an up-down sorting engine by actually designing its basic components, the compare/steer unit, and understanding its overall circuit organization and operations.

Operation of the Sorter

The sorting engine consists of a series of compare/steer circuits. The total number of compare/steer circuits determines the maximum number of values that can be compared. Figure 6-6 represents pictorially the operation of a single compare/steer circuit. The arrows indicate a vertical and lateral movement within the circuit for the data entering and exiting the circuit. The downward solid line represents the input movement. The upward dashed line represents the output movement. The double lines are channels for routing the results of the comparator and the storage cells. Blanks A, B, C, and D represent cells in the circuit. Cells A and B are the input cells; cells C and D are storage element cells that are initialized to be infinitely large.

The engine operates in two phases: input unsorted values and output sorted values. During the input phase, cell A receives an external input and cell B receives an internal input from cell C. The value in cell C represents the minimum value as a result of the comparison of cells A and B; cell D represents the maximum value of the comparison. During the output phase, cell A receives as input cell D, and cell B receives an external value. Cell A's external input is either a new value to the engine or a value from the previous compare/steer unit. Cell B's external input is from the immediately succeeding compare/steer unit. During the input phase, cell D's output serves as an input to the succeeding compare/steer unit. During the output phase, cell C's output serves as an input to the preceding compare/steer unit or output of the sorting engine. Figure 6-6 illustrates the operations of the two phases for a compare/steer unit. Figure 6-7 demonstrates a 6-stage sorting engine.

The sorting engine attributes are:

1. the sorting of n values requires only n/2 comparisons (integer division) and
2. the input/output speed of the engine determines the real time of the sort.

Figure 6-8 is a histogram of the sorting operations during the input/output phase of the sorting of six numbers. Note that the values inputted to the engine are effectively sorted immediately. The complete time required to perform the function equals the time required to input and output the values. The circuit requires the engine compare/steer circuits to be initialized to a value larger than a value to be compared. For the histogram, infinity is an appropriate value. You should also note that an infinity value is required as input to terminal node of the engine to output the values. This infinity value must not change during outputting. If the value changes, the value will be compared and steered, possibly destroying the sort. The engine can be initialized by performing an output function for the size of the sorting engine. Once initialized, the normal outputting of the sorting engine reinitiates each compare/steer circuit.

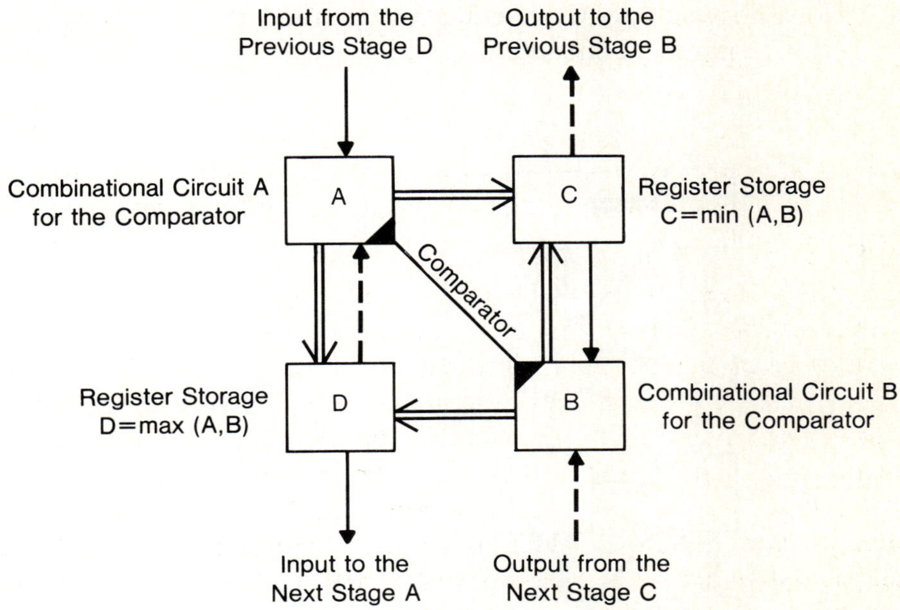

Figure 6-6. *A Compare/Steer Unit Diagram (Adapted from Figure 1, [Reference 1 of VLSI Systems in the Bibliography], D. T. Lee, et al.,* **IEEE Transactions on Computers**, June, 1981, © IEEE 1981)

Design, with pencil and paper only, the circuit of the sort engine just described.

Note that C and D are registers in the compare/steer unit. A sample solution is shown in Figure 6-9 for 4-bit registers. Since A and B are actually input points to the compare/steer unit, there is no need to use registers for them.

Questions:

1. We need to build a sorting engine to sort 99 binary numbers (between 0 and 999) in 200 ms. List 1) the minimum number of comparators required, 2) the minimum speed of the clock for the VLSI, 3) the minimum number of bits of storage needed, 4) the minimum size of the comparator, 5) the minimum number of 2x1 "multiplexer"-type of circuits required, and 6) the number of comparisons that happens for each clock time in the VLSI sorting engine.

2. Draw the circuitry that is required for initializing the VLSI sorting engine.
3. How do you prove that the circuitry of the project can indeed sort things correctly for every input sequence? Validate that it works for the input sequences 5, 4, 7, 6, 1, and 8.
4. How do you modify the design so that the VLSI will sort positive numbers out in a *descending* order?

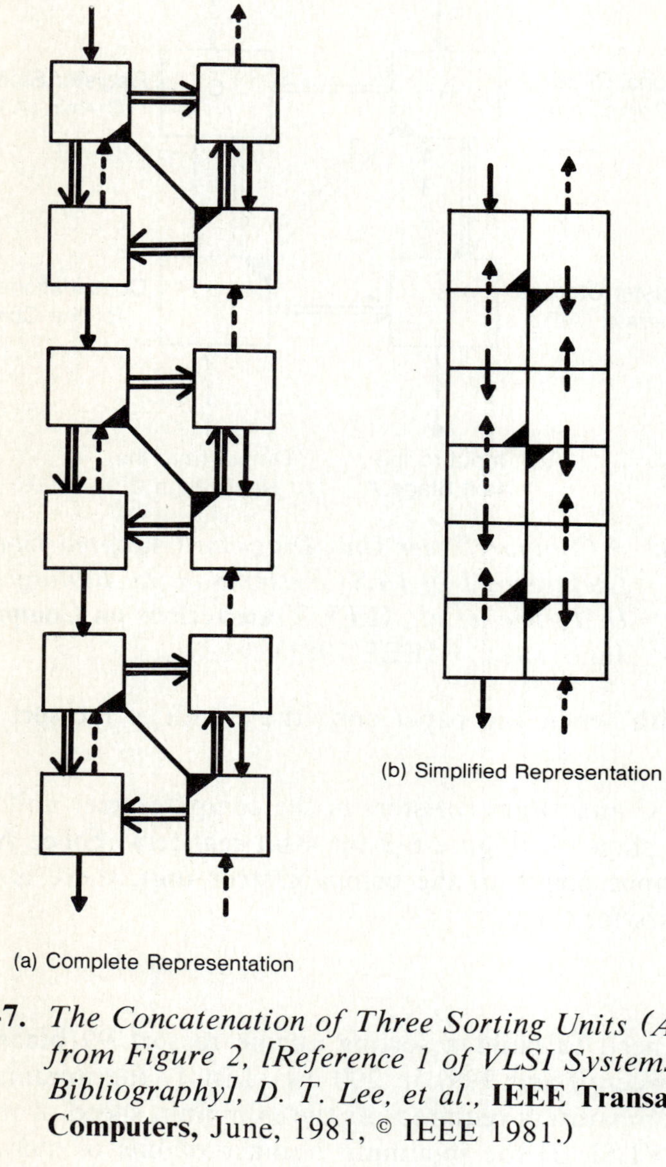

Figure 6-7. *The Concatenation of Three Sorting Units (Adapted from Figure 2, [Reference 1 of VLSI Systems in the Bibliography], D. T. Lee, et al.,* **IEEE Transactions on Computers,** June, 1981, © IEEE 1981.)

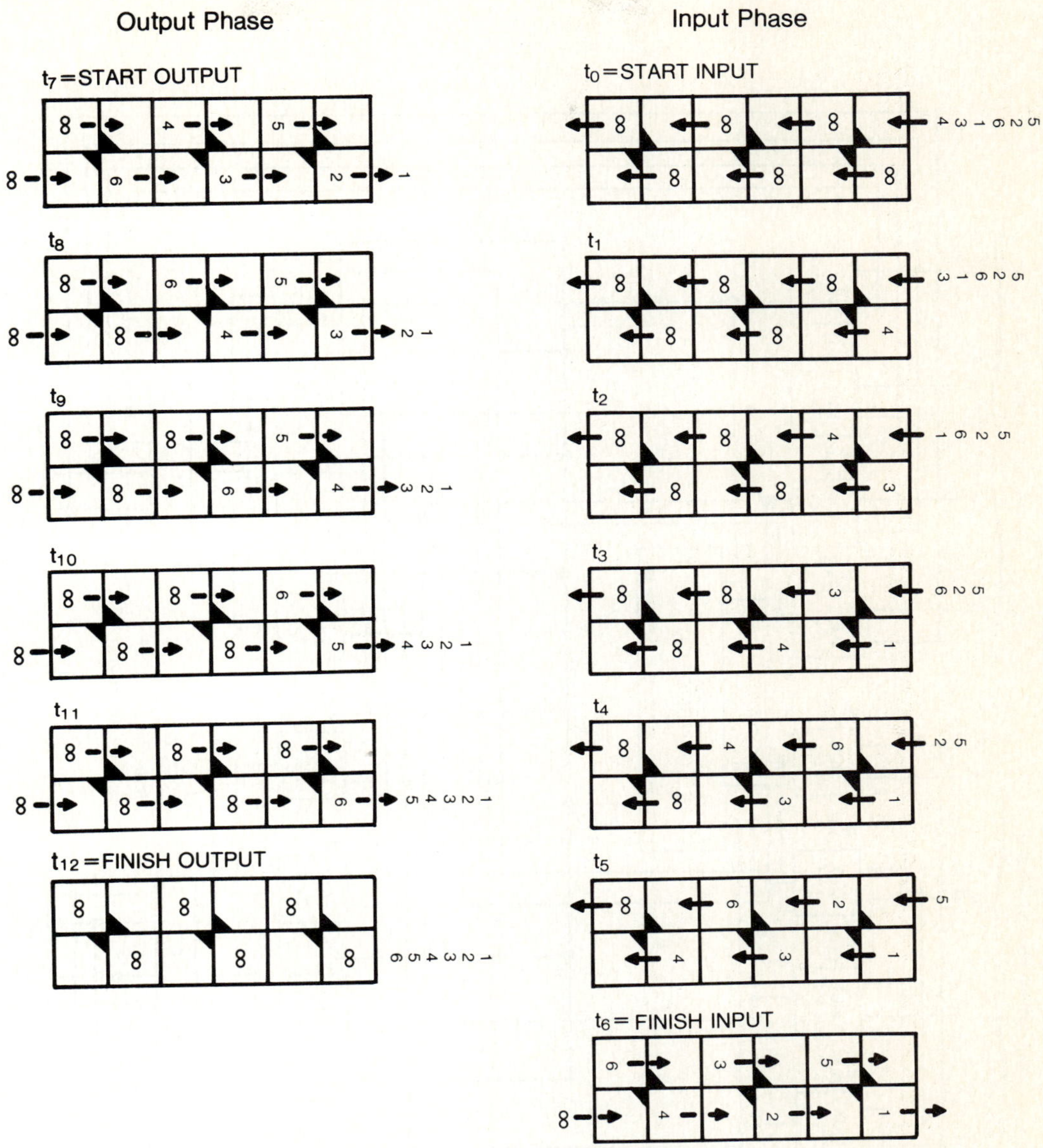

Figure 6-8. *A Histogram of the Operation of the Up/Down Sorter, With Input (4, 3, 1, 6, 2, 5) and Output (1, 2, 3, 4, 5, 6) (Adapted from Figure 3, [Reference 1 of VLSI Systems in the Bibliography], D. T. Lee, et al.,* **IEEE Transactions on Computers,** *June, 1981, © IEEE 1981.)*

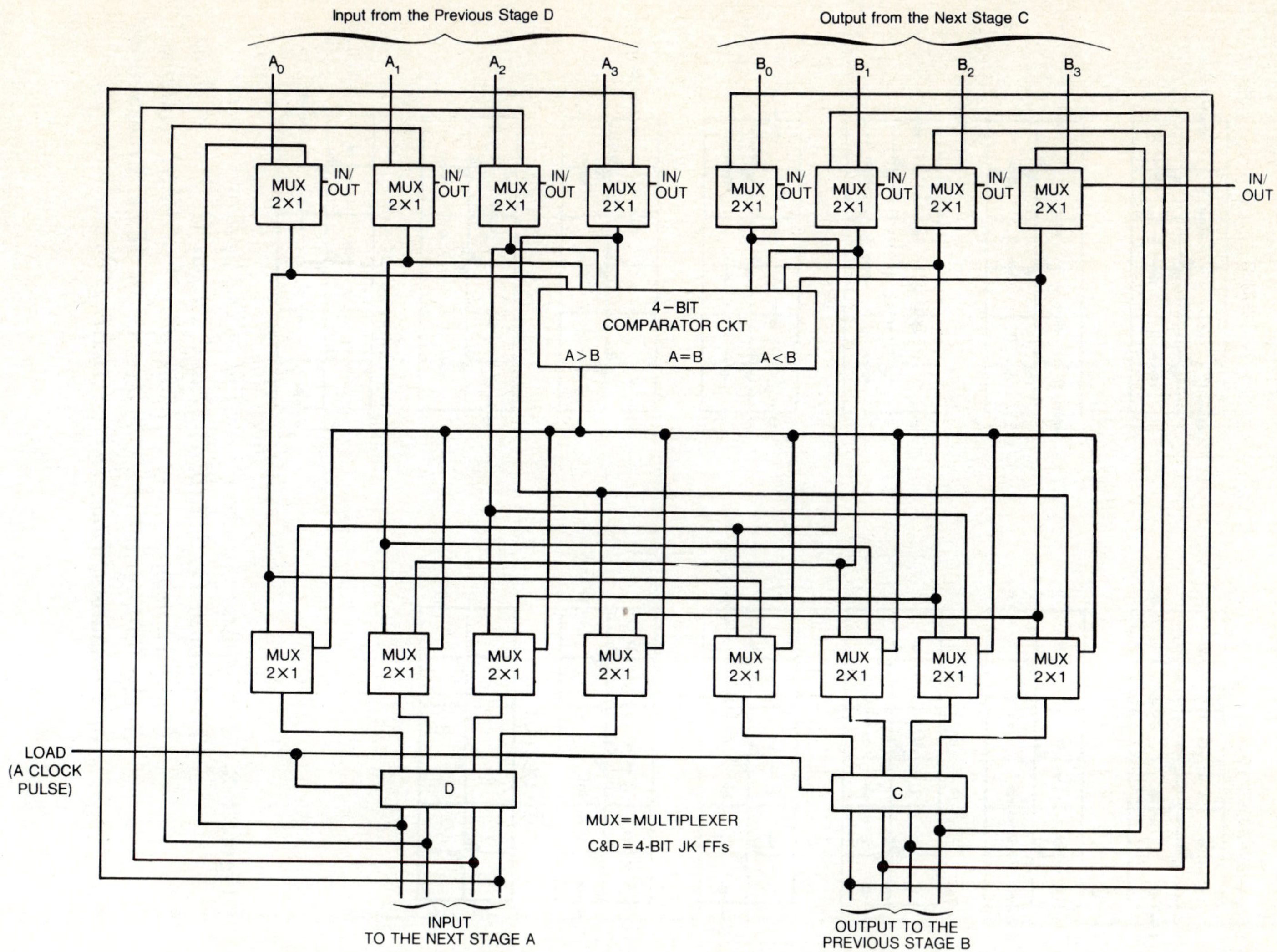

Figure 6-9. *The Circuit Diagram of the Compare/Steer Unit*

6.3 Project 3 - The Interface Design of a Local Area Network

Besides the sorting engine, another interesting area for VLSI application is telecommunications. Terminal/host computer interface circuits with communication networks are of particular interest. This project is adopted from the initial "Register Insertion Ring" designed by Ming T. Liu and C. Reams (see Reference 1 of Computer Communications in the Bibliography) and is limited to the design of the register insertion ring node only.

6.3.1 Ring Network Overview

A ring network is a high-speed local area network that can transmit data with several million bits per second. The register insertion ring is a network connected by a loop topology as shown in Figure 6-10. The ring is connected by the nodes, called register insertion nodes, and a station can be attached to a node to communicate with any other station attached to a node in the ring. The stations can be either host computers, terminals, or peripheral devices such as printers.

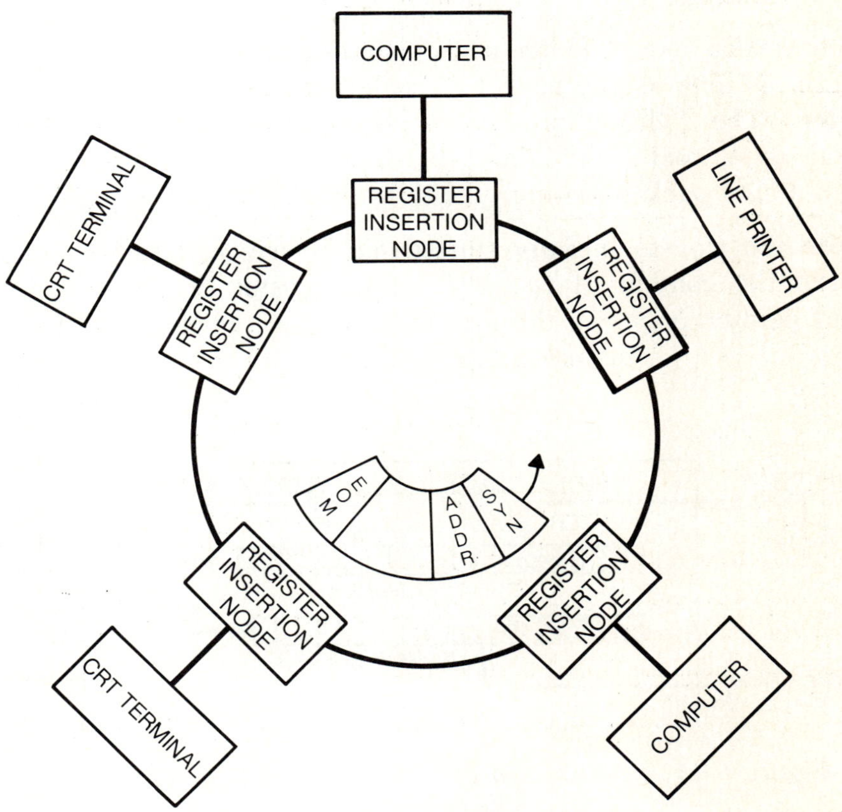

Figure 6-10. *Ring Network*

A register insertion ring allows more than one station in the network to transmit variable length packets at the same time by having a delay buffer at each station of the ring network. Many network services such as file transfers between stations can be provided by using the ring network or other underlying data transfer facilities (see Reference 4 of Computer Communications in the Bibliography).

6.3.2 Design Description

Each node in the network consists of the following basic components:

1. An input-shift register called a Delay Buffer (DB) that is the same size for each station and is large enough to hold all bits of the maximum packet size sent on the ring.
2. An Output Buffer (OB) that stores the output packets from the local station. The packet sent by the station can be variable length but no longer than the size of the OB. Both DB and OB are 256 bytes in this design example.
3. An input pointer and an output pointer.
4. A switch that can choose between outputting the input-shift register or OB.

Each station can send its packet simultaneously to any destination station as long as the packet is no larger than the size of the delay buffer. The collision problem is solved since the ring maintains a serial data path with the DB "breaking" the physical ring yet maintaining packet continuity with the logical ring (see Figure 6-11).

When the ring is initially started, all interface switches are set to output the content of the input-shift register or delay buffer, the input pointer points to the first stage of the register, and (assuming

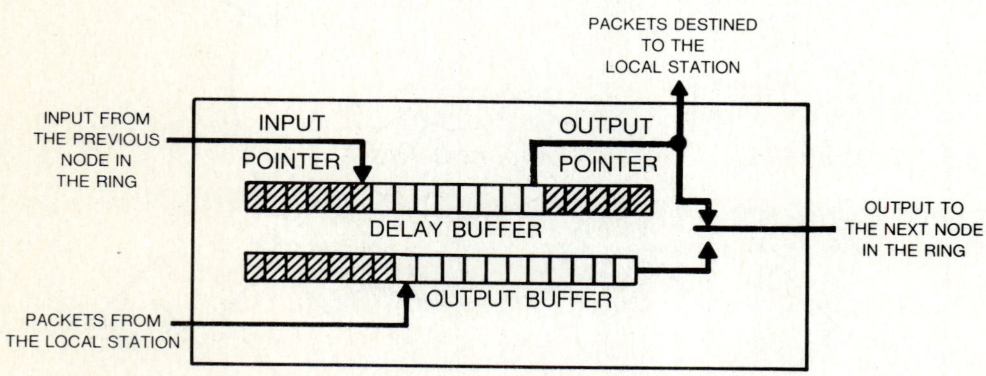

Figure 6-11. *Register Insertion Node (Logical View) (Adapted from Figure 7-14, [Reference 2 of Computer Communications in the Bibliography, A. S. Tanenbaum,* **Computer Networks**, *1981, Prentice-Hall)*

the interfaces have no output packets to send) the OB counter is set to zero.

Let us first assume that a station has a packet to output. Since the OB is empty, the station transfers the packet from the station's internal memory to the OB. The interface compares the size of the packet with the amount of space remaining empty in the input buffer. As long as there are more empty positions in the input buffer than the number of bits in the packet to be sent out of the station, the switch can be changed to output the message packet from the station. On completing outputting the packet, the switch is returned to output from the input-shift register. This way, we can be sure that the interface will not lose any incoming information while performing the output of its own station.

To determine what action is to be made on a packet arriving into its input-shift register, the packet's address is checked. If a packet is addressed to a local station, the node "removes" the packet and may generate an acknowledgment to the sender. The DB acts like a shift register to output to the next ring node if the packet is not addressed to its station. The DB may have input and output at the same time, or have output only during input gaps, or have input only during the output mode for OB. In Liu's paper, he describes an actual implementation of this circuit. We modify the circuit so as to use the notation conventions of this book.

It should be noted that the register insertion node has two functions: ring interface and station interface. It is the function of the ring interface to transmit the selected output source into the ring. It is the function of the station interface to put its output packets into the OB. The synchronization between the ring and station interfaces is through the RDY and Mode (M) FFs.

The following describes the design by micro-operations. Figures 6-12 and 6-13 describe the operations of the ring interface design and the circuit for its implementation.

6.3.2.1 Ring Node Register Component Description

- An 8-bit Input Pointer (IP) register that points at the DB input position for the data from the previous node in the ring.
- An 8-bit Output Pointer (OP) register that points at the DB output position to output the data to the next node in the ring.
- An 8-bit Transmit Counter (TC) register that counts the number of characters remaining in the OB.
- A 9-bit Delay Counter (DC) register that is always set to the sum of the TC and the input message length remaining in the DB.

- An 8-bit Received Input (RI) register that stores the current incoming data from the ring.
- An 8-bit Transmitted Output (TO) register that stores the current data transmitted out to the loop from the DB.
- A Ready (RDY) FF that indicates that an output message is ready in the OB when it is 1, and indicates that a new message can be put in the OB when it is 0.
- A transmit/relay M FF that indicates whether the node is transmitting data from the DB or the OB.

6.3.2.2 Ring Node Circuit Descriptions

Initialize Node

Initialize: $RDY \leftarrow 0$, $TC \leftarrow 0$, $DC \leftarrow 0$, $M \leftarrow 0$,
$IP \leftarrow 0$, $OP \leftarrow 0$, $TO \leftarrow$ 'SYN', $DB(0) \leftarrow$ 'SYN'

Ring Interface

C_1: RI ← INPUT CHARACTER, DB(IP) ← INPUT CHARACTER
if $(M = 0)$ *AND* packet_address = local_station,
then *STATION* ← $DB(OP)$ else *OUTPUT* ← $DB(OP)$
if $(M = 0)$ then $TO \leftarrow DB(OP)$
if $(M = 1)$ then *OUTPUT* ← $OB(TC)$

Note: The OB acts like a FIFO memory for outputting the node-generated message to the loop data out bus and to the TO register. The first character into the OB is the first character outputted. As a character is removed from the buffer, the remaining characters in the buffer are shifted down (towards the output) one position. One character is removed each C_2 time.

C_2: if $(RI \neq$ 'SYN'$)$ then $IP \leftarrow IP + 1$
if $(RI =$ 'SYN' and $DC \neq 0)$ then $DC \leftarrow DC - 1$
if $(M = 1)$ then $TC \leftarrow TC - 1$
if $(M = 0$ and $OC \neq IC)$ then $OP \leftarrow OP + 1$

Station Interface

$RDY' \cdot NEW\_MSG \cdot C_1$:	$OB(TC) \leftarrow$ station input
$RDY' \cdot NEW\_MSG \cdot C_2$:	$TC \leftarrow TC + 1$, $DC \leftarrow DC + 1$ if (NEW_MSG_LOADED) then $RDY \leftarrow 1$, $NEW\_MSG \leftarrow 0$

The NEW_MSG flag is on when a message at the station is waiting to be loaded into the OB. If no other message is already loaded in the OB, the station loads the OB with all characters of the message to be outputted. The buffer (OB) acts like a FIFO memory. As the characters are inputted, each character is pushed down to the correct position in the FIFO. After all of the characters of the new message are loaded into the OB, NEW_MSG_LOADED is set to 1 and NEW_MSG is set to 0. The RDY flag is then set to 1 to indicate that a message from the local station is ready for output.

The TC is always set to the message length remaining in the OB and the DC is always set to the sum of the TC and the input message length remaining in the DB.

Synchronization Station and Ring Interfaces

if $((RDY = 1)$ and $TO = ('EOM' + 'SYN')$ and $DC < 256)$ then $M \leftarrow 1$
if $(TC = 0)$ then $RDY \leftarrow 0, M \leftarrow 0$

Questions:

1. Can the size of the DB be smaller than the size of the OB? Why?

2. Since the channel between any two nodes can only send one packet at a time, how can collision be avoided if several stations in a register insertion network send packets at the same time?

3. Complete the design for the output buffer circuit, the output selection circuit, and the delay buffer counter circuit in Figure 6-13.

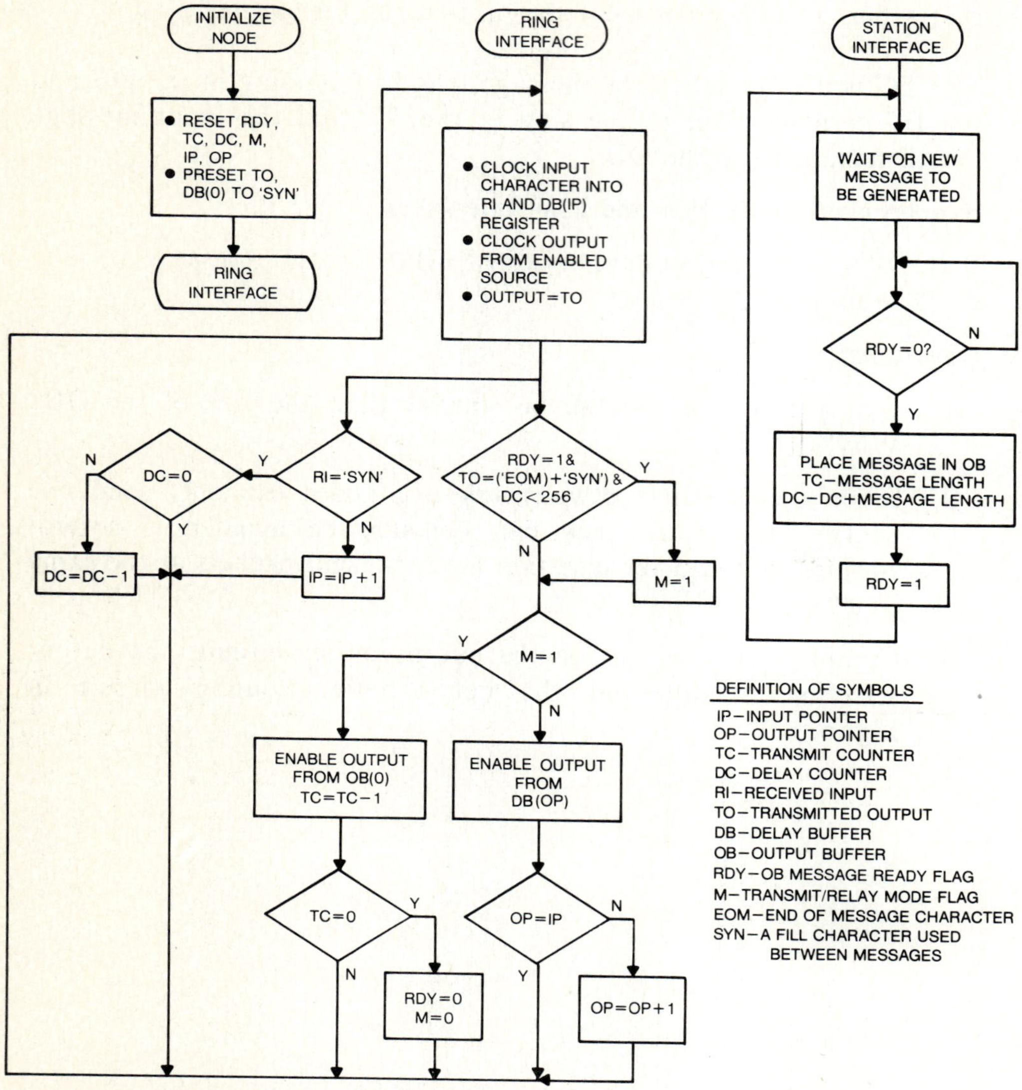

Figure 6-12. *Operation of Loop Transmitter Using Shift-Register Insertion Mechanism (Adapted from Figure 4, [Reference 1 of Computer Communications in the Bibliography], M. T. Liu and C. C. Reames,* **Proc. International Computer Symposium,** *1975.)*

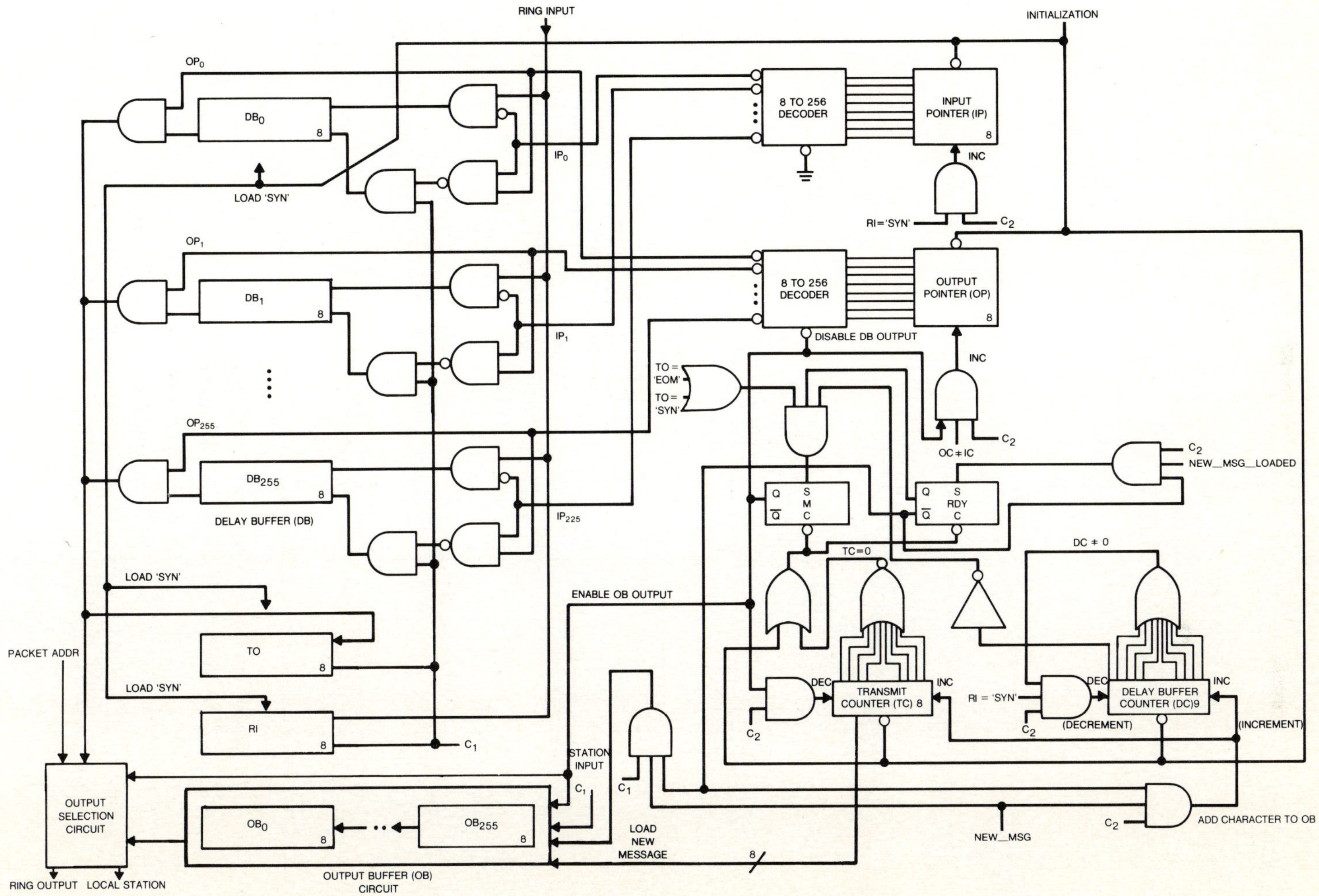

Figure 6-13. *Hardware Implementation of Shift-Register Insertion Transmission Mechanism (Adapted from Figure 3, Reference 1 of Computer Communications in the Bibliography], M. T. Liu)*

Appendix A
Combinatorial Circuits

The necessary basis for the design of logic circuits is a working knowledge of boolean algebra, logic gates, and combinational circuits. In this section, we give a brief overview of these topics to provide such a basis. If you feel the need to know more about these topics, you should then study them from any standard logic design textbook.

Boolean Algebra

In boolean algebra, we combine variables into expressions with operators that obey certain algebra laws. The boolean variables denoted by letters of the alphabet, A, B, C, etc., may have one of the two binary values only: 1 or 0. These are sometimes read as true and false or as high and low. There are three basic boolean operators: OR, AND, and NOT. The OR operator is written as +. $A + B$ is true if either A is true or B is true, and false otherwise. The AND operator is written as "·", or omitted when the variables are known to be denoted by single letters. $A \cdot B$ (or AB) is true if A and B are both true, and false otherwise. The NOT (complementing or negation) operator is written as a bar over its arguments or a ' after it. $\overline{A}$ is true if A is false, and false otherwise.

There are several important laws we should remember for boolean algebra. Both AND and OR operators are commutative, associative, and distributive, i.e., $A + B = B + A$, $A + (B + C) = (A + B) + C$, and $A(B + C) = AB + AC$. Another important law is DeMorgan's law: the complement of a boolean expression can be derived directly by replacing each variable by its complement in the corresponding dual expression. For example, $\overline{A + B} = \overline{A} \cdot \overline{B}$, $\overline{AB} = \overline{A} + \overline{B}$, and $\overline{A + BC} = \overline{A} \cdot \overline{BC} = \overline{A} \cdot (\overline{B} + \overline{C})$.

Logic Gates

The boolean algebra operators can be implemented by hardware logic gates. Gates are the basic building block elements of all IC digital circuitry. The most complicated computer is nothing but gates combined to perform complex functions. The following diagram shows the five gates that are used in this book: AND, OR, NOT, NAND, and EXCLUSIVE-OR.

Gates are simple devices that have exact rules which define what comes out of the device for every possible combination of what goes

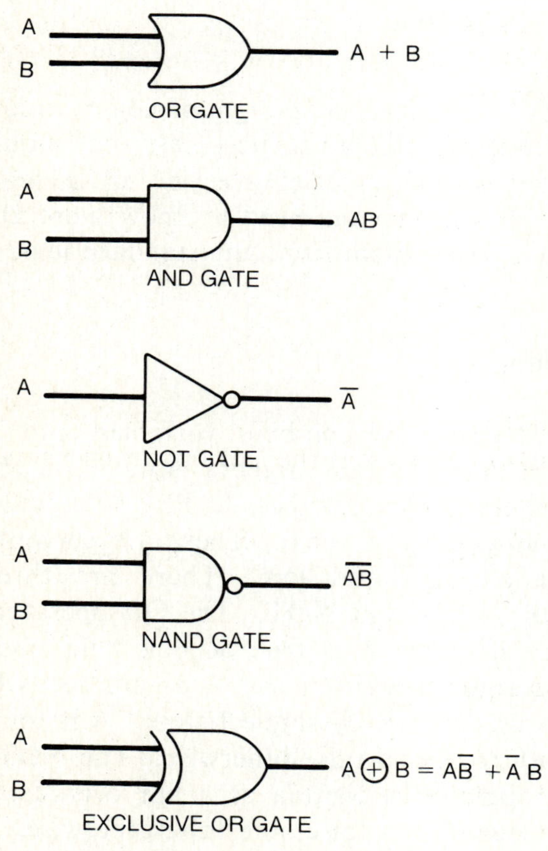

Logic Gates

in. What goes into (input) and comes out of (output) gates are either logic 1 or logic 0.

The AND, OR, and NOT gates obey the same laws as the AND, OR, and NOT operators in boolean algebra. The definition of the OR gate is: the output is 1 if the least one input to the OR gate is 1. If all of the input are 0s, then the output is 0. The definition of the AND gate is: the output is 1 if all the inputs to an AND gate are 1s. If any input is not 1, then the output is 0. The INVERTER changes the state of its input. If the input is 1 then the output is 0 and if the input is 0 then the output is 1. For the NAND gate, the output is 0 if all inputs are 1s. If any of the input is 0 then the output is 1.

The NAND gate is the most important one in implementing the experiments of the book. It is actually a combination of the AND gate and the inverter. The inverter function is often called the NOT function, and NAND is a contraction of the names of its two component gates, NOT-AND. Most logic lines offer a variety of NAND, AND, OR, and NOT gates. Because NAND gates can be used to construct any other logic gate, NAND is the most basic logic gate offered. You should review the technique for constructing various logic gates from the NAND gate.

The EXCLUSIVE-OR gate is defined: the output is 0 if the inputs are the same. If the inputs are not the same then the output is 1. The gate is used to compare signals to see if they are different.

A product term is represented by the AND of the appropriate literals and a sum term is represented by the OR of the appropriate literals. A function is often represented in the sum of products form as will be illustrated in the next section.

Questions:

1. Implement the NOT, AND, and OR operations using NAND gates. Also, implement the OR operation using NOR gates.
2. Verify that we can simply change all AND-OR gates of the AND-OR gate approach to NAND gates and still implement the same function.

Combinational Logic Circuit

Notice that every boolean function can be implemented by several combinational logic circuits. For example, $f = A + AB + C$ can be realized by two gates:

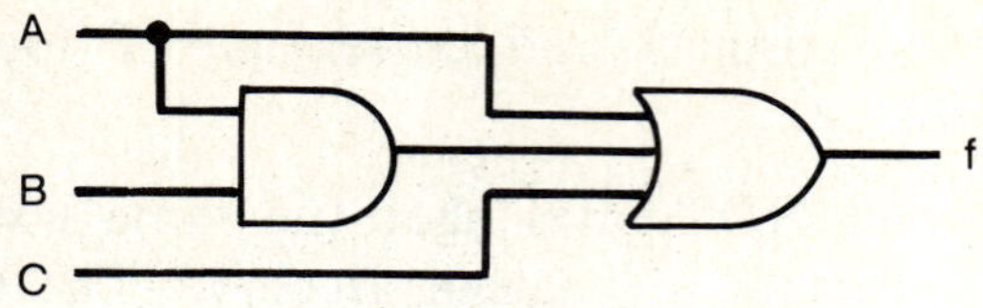

However, since f = A + AB + C = A + C, the same function can also be implemented by one OR gate.

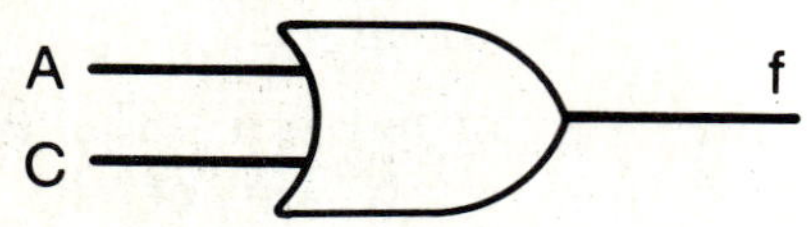

In order to minimize the use of IC chips for implementing a boolean function, we would like to use the minimum number of gates in general. How can we determine if we are using the smallest circuit in implementing our function?

The three steps for deriving the smallest combinational circuit for a boolean function are as follows:

1. Multiply out the given boolean function and convert it into its 2-level sum-of-products form.

2. Use a Karnaugh map to systematically simplify the function so as to derive the minimal sum-of-products expression of the boolean function. (The details of the use of Karnaugh maps are beyond the scope of this book. You should consult any logic design textbook for the notations and the use of Karnaugh maps.)

3. Implement the sum-of-product expression by AND-OR gates or NAND gates (INVERTERS do not count). The AND-OR gates approach uses an AND gate for each product and then uses an OR gate in the second level with all outputs of the AND gates as inputs. To use NAND gates to implement a function, we can simply change all AND-OR gates of the AND-OR gate approach to NAND gates. The NAND-gate-only approach is very important for the experiments in this book since you may not have an AND-OR gate in your parts kit.

An example of deriving the minimal NAND gate implementation for a 3-input majority function is shown in Experiment 1.

Appendix B

Circuit Synthesis and Analysis

SYNTHESIS

Circuit synthesis refers to the procedure of designing a circuit from a verbal description. Circuit analysis refers to the procedure of analyzing a circuit already built or designed. To design a new circuit so as to implement a desired function requires circuit synthesis and to study a given circuit so as to understand its functions and use requires circuit analysis. The steps of circuit synthesis are:

1. study the given verbal description,
2. develop a state diagram that describes the transitions of the FF content,
3. develop a state table that is a table representation of the state diagram,
4. develop an excitation table that describes the desired input values to the FFs to achieve the state transitions,
5. develop an excitation and output function, and
6. produce the resulting sequential circuit.

Synthesis Example

The following is a synthesis example of BCD counter circuit (see Experiment 5).

Verbal Description

A BCD counter uses 4 bits to represent the 0 to 9 decimal digits and counts from 0 to 9, and then returns to 0 when the input is "1" and clock is 1. The count sequence is (0, 1, 2, 3, 4, 5, 6, 7, 8, 9, 0, 1...9). A reset pulse initializes the counter to '0000'.

State Table/Transition Table/Excitation Table

Figure B-1 is the state diagram for the BCD counter. From the state diagram, the state table can be created (see Table A). It should be noted at this time that when the input is "0" there is no change in state. Additionally, the count is from "0000" to "1001"; counts from "1010" to "1111" can never happen and can be treated as "do not care" states.

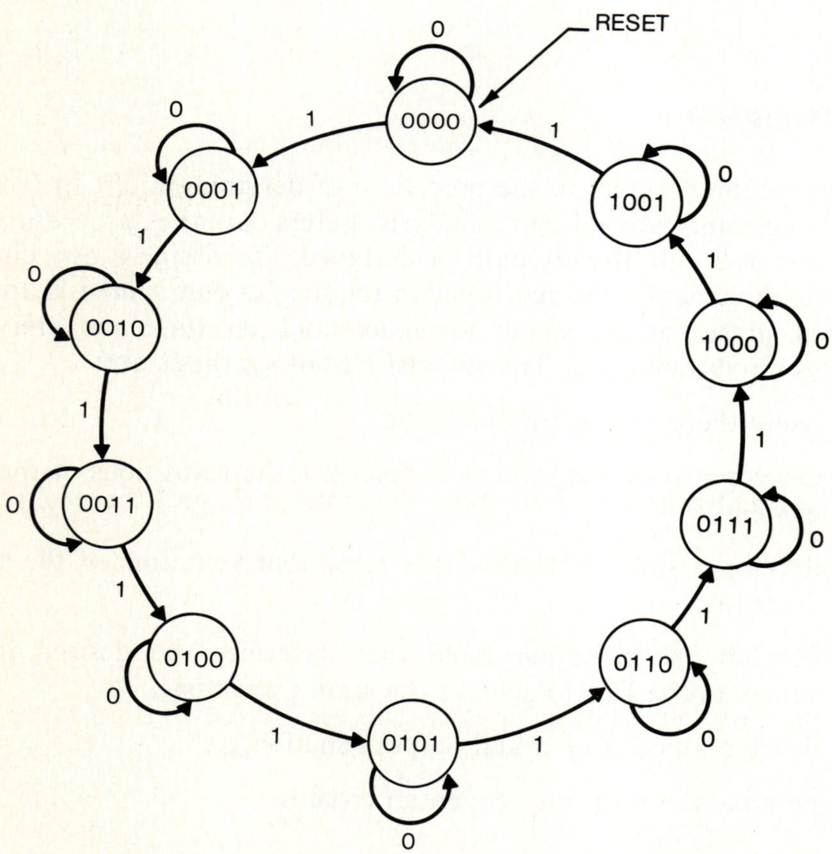

Figure B-1. *The State Diagram of a BCD Counter*

Table A. *The State/Excitation Table of the BCD Counter*

STATE TABLE						
INPUT	CURRENT STATE	NEXT STATE	FLIP-FLOP INPUTS			
X	D C B A	D C B A	J_d K_d	J_c K_c	J_b K_b	J_a K_a
0	0 0 0 0	0 0 0 0	0 x	0 x	0 x	0 x
0	0 0 0 1	0 0 0 1	0 x	0 x	0 x	x 0
0	0 0 1 0	0 0 1 0	0 x	0 x	x 0	0 x
0	0 0 1 1	0 0 1 1	0 x	0 x	x 0	x 0
0	0 1 0 0	0 1 0 0	0 x	x 0	0 x	0 x
0	0 1 0 1	0 1 0 1	0 x	x 0	0 x	x 0
0	0 1 1 0	0 1 1 0	0 x	x 0	x 0	0 x
0	0 1 1 1	0 1 1 1	0 x	x 0	x 0	x 0
0	1 0 0 0	1 0 0 0	x 0	0 x	0 x	0 x
0	1 0 0 1	1 0 0 1	x 0	0 x	0 x	x 0
0	1 0 1 0	x x x x	x x	x x	x x	x x
0	1 0 1 1	x x x x	x x	x x	x x	x x
0	1 1 0 0	x x x x	x x	x x	x x	x x
0	1 1 0 1	x x x x	x x	x x	x x	x x
0	1 1 1 0	x x x x	x x	x x	x x	x x
0	1 1 1 1	x x x x	x x	x x	x x	x x
1	0 0 0 0	0 0 0 1	0 x	0 x	0 x	1 x
1	0 0 0 1	0 0 1 0	0 x	0 x	1 x	x 1
1	0 0 1 0	0 0 1 1	0 x	0 x	x 0	1 x
1	0 0 1 1	0 1 0 0	0 x	1 x	x 1	x 1
1	0 1 0 0	0 1 0 1	0 x	x 0	0 x	1 x
1	0 1 0 1	0 1 1 0	0 x	x 0	1 x	x 1
1	0 1 1 0	0 1 1 1	0 x	x 0	x 0	1 x
1	0 1 1 1	1 0 0 0	1 x	x 1	x 1	x 1
1	1 0 0 0	1 0 0 1	x 0	0 x	0 x	1 x
1	1 0 0 1	0 0 0 0	x 1	0 x	0 x	x 1
1	1 0 1 0	x x x x	x x	x x	x x	x x
1	1 0 1 1	x x x x	x x	x x	x x	x x
1	1 1 0 0	x x x x	x x	x x	x x	x x
1	1 1 0 1	x x x x	x x	x x	x x	x x
1	1 1 1 0	x x x x	x x	x x	x x	x x
1	1 1 1 1	x x x x	x x	x x	x x	x x
INPUT TO COMBINATION CIRCUITS			EXCITATION TABLE			

It is a normal practice to combine the state table and the excitation table as shown in Table A.

Excitation Functions and Circuits

From the excitation table, Karnaugh maps can be created to derive excitation and output functions as shown in Figure B-2. From this data, the circuit can be produced. See Figure 4-21 in Experiment 5 BCD counter for the resulting circuit. Adding the input x to the circuits in Figure 4-21 would make the circuit asynchronous per this circuit description. Note the J and K inputs would also require an "AND" input with input x.

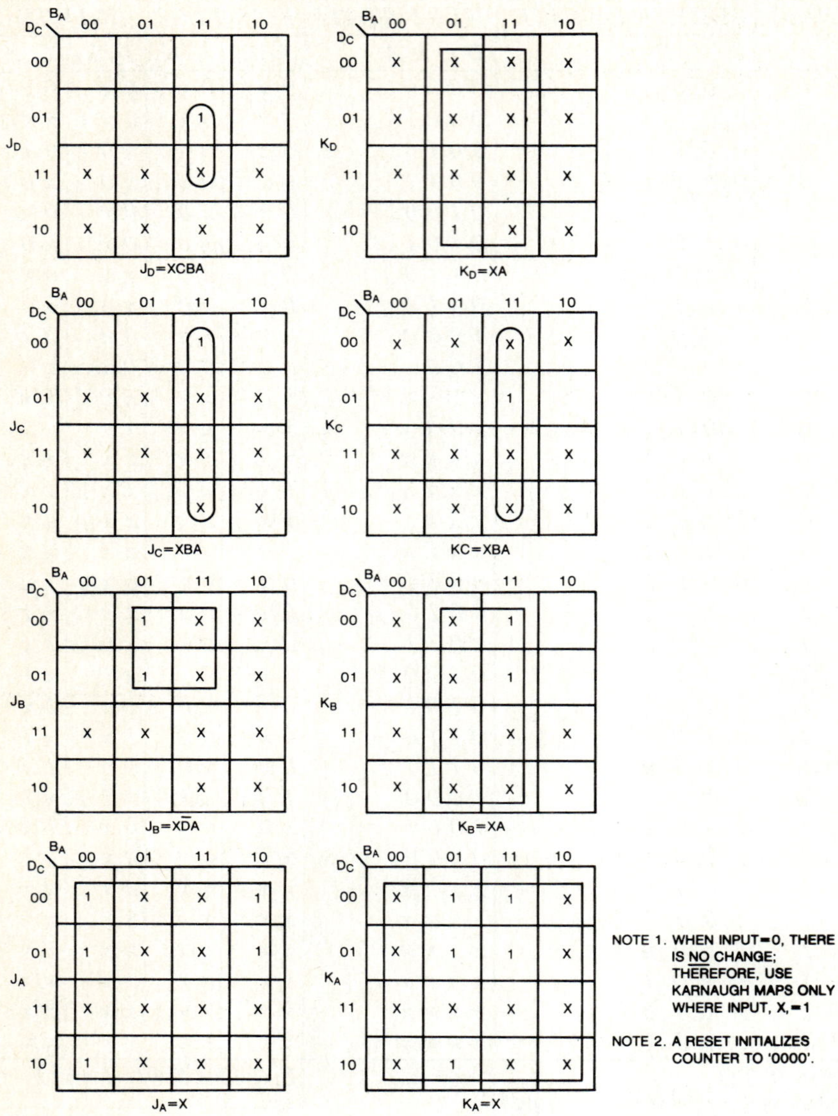

Figure B-2. *Karnaugh Maps of the BCD Counter Circuit*

ANALYSIS

Circuit analysis involves the same steps but in reverse order. Thus, the steps in circuit analysis are:

1. a given sequential circuit,
2. develop an excitation and output specification,
3. develop an excitation table,

4. develop a state table,
5. develop a state diagram, and
6. give verbal description of the circuit.

As in circuit synthesis, Steps 3 and 4 are usually combined.

Analysis Example:

The following description is an analysis of a 2-stage up counter. See Figure B-3.

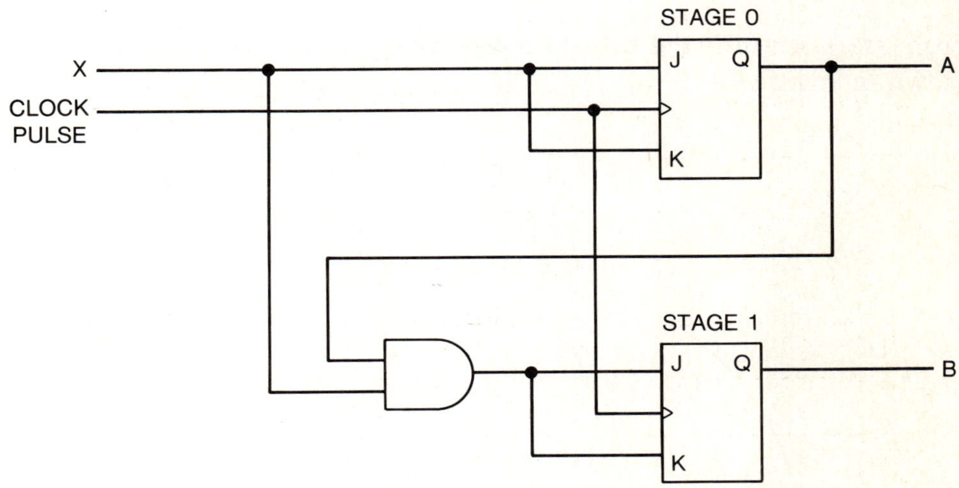

Figure B-3. *Two-Stage Up Counter Circuit*

Circuit

From the circuit in Figure B-3, we can prepare a descriptive "excitation and output function" specification.

Excitation and Output Functions

$$Q_A = \overline{Q_A} \cdot X$$

$$\overline{Q_A} = Q_A \cdot X$$

$$Q_B = Q_A \cdot X \cdot \overline{Q_B}$$

$$\overline{Q_B} = Q_A \cdot X \cdot Q_B$$

Excitation Table/State Table

From the output functions the excitation and state tables can be created.

Present State Input			*Next State*		*FF Inputs*	
B	*A*	*X*	*B*	*A*	J_b K_b	J_a K_a
0	*0*	*0*	*0*	*0*	*0 X*	*0 X*
0	*0*	*1*	*0*	*1*	*0 X*	*1 X*
0	*1*	*0*	*0*	*1*	*0 X*	*X 0*
0	*1*	*1*	*1*	*0*	*1 X*	*X 1*
1	*0*	*0*	*1*	*0*	*X 0*	*0 X*
1	*0*	*1*	*1*	*1*	*X 0*	*1 X*
1	*1*	*0*	*1*	*1*	*X 0*	*X 0*
1	*1*	*1*	*0*	*0*	*X 1*	*X 1*

State Diagram

From state parts of the tables, a state diagram can be constructed as shown in Figure B-4.

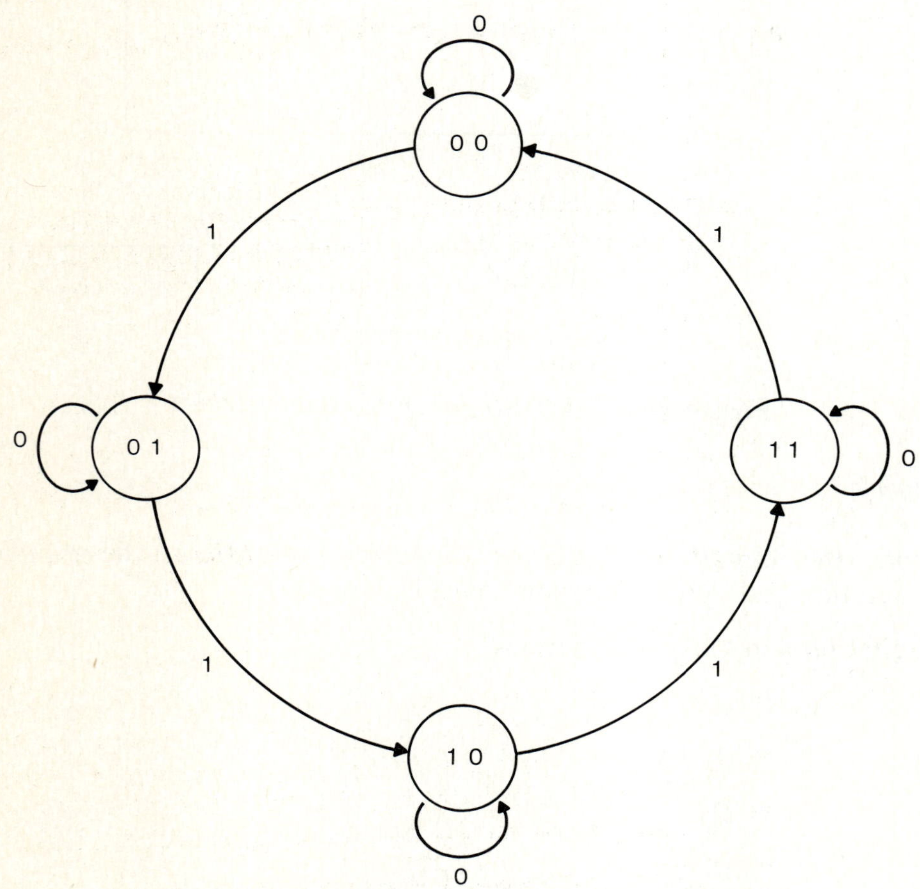

Figure B-4. *The State Diagram of the Up Counter*

Verbal Description

From the state diagram, a verbal description can be generated. For the circuit being analyzed, the following circuit specification can be made:

1. The circuit has four unique states: "00", "01", "10", and "11".
2. It takes an input equal to "1" to cause a state change.
3. The state changes are in one direction only (up) and resemble binary digits.
4. The circuit is a 2-stage up counter.

Appendix C

A Hardware Description Language

Introduction

There are many different hardware description languages available to describe hardware design. This book uses a hardware language known as RT language (similar to the language used in Reference 1 of Computer Design) to describe the various circuit operations. RT language expresses data transfer between registers. The transfer can be a direct transfer without any arithmetic or logical operation performed on the data or a transfer with an arithmetic or a logical operation performed on the data, etc. In this book, RT language can also be used to define control timing sequences. Before explaining the syntax of the RT language, let us first discuss the notation used to express hardware micro-operations.

We use capital letters to denote registers (such as MAR) and memories (such as M). To denote a bit in a register, we put a subscript representing the bit position beside the register. For example, A_2 represents the second bit of Register A. We can describe a portion of a register inside parentheses. For example, A(1-3) denotes the A_1, A_2, and A_3; the MBR(AD) may denote the address portion of the MBR register.

Micro-Operations

The transferring of data, performing arithmetic or logical operations on data, and changing from one control state to another are considered micro-operations. These micro-operations can be expressed symbolically or in RT language statements.

The parallel transfer of data from a register A, containing 8 bits, to a register B, also an 8-bit register, can symbolically be expressed as B ← A or B(1-8) ← A(1-8).

The "←" indicates the transfer of data or information. The serial transfer of data (1 bit at a time) can also be expressed symbolically; this is expressed as follows:

$$B_1 \leftarrow B_2, B_2 \leftarrow B_3 \cdots B_7 \leftarrow B_8, B_8 \leftarrow A_1,$$

$$A_1 \leftarrow A_2, A_2 \leftarrow A_3 \cdots A_7 \leftarrow A_8, A_8 \leftarrow \text{serial\_input}$$

The simple addition of register A's data with a data in B register with the result transferred to B register is expressed as follows:

$$B \leftarrow A + B$$

RT Language

RT language is a rather simple language to describe hardware micro-operations. Like computer programming languages, the statement sequence generally defines the sequence to be executed.

The basic syntax for the RT language is best described by an example:

$$P\text{: expression}$$

"P" represents a control function that will be designed in hardware or firmware. The colon marks the end of the control functions name. When the control function P is logically true, the expression is executed. The expression is a micro-operation or a series of micro-operations separated by commas.

Examples of RT language statements are as follows:

$$P\text{: } B \leftarrow A + B$$
$$Pt_1 \; B \leftarrow A \cdot B$$

An example of RT language used to express a control function and modifying the sequence of steps is as follows:

t_1 : $A\ (0-3) \leftarrow 0,\ FLAG \leftarrow 0$

t_2 : if $(\ \overline{READY}\ )$ then $t_2 \leftarrow 1$

t_3 : $FLAG \leftarrow 1$

t_4 : $A\ (0-3) \leftarrow A\ (0-3) \oplus IN\ (0-3)$

t_5 : $FLAG \leftarrow 1$

t_6 : $t_2 \leftarrow 1$

where "FLAG" and "READY" are FFs, Step 6 (t_6) is an unconditional branch to t_2, and Step 2 (t_2) is a conditional branch.

Once a circuit design is described precisely in RT language, we can systematically implement the design by using IC chips. Experiment 12 of the book has an example of producing a table of micro-operations by registers from the RT language to help derive the circuit for the design.

Appendix D

Pin Layouts and Function Tables for IC Chips Used in the Book

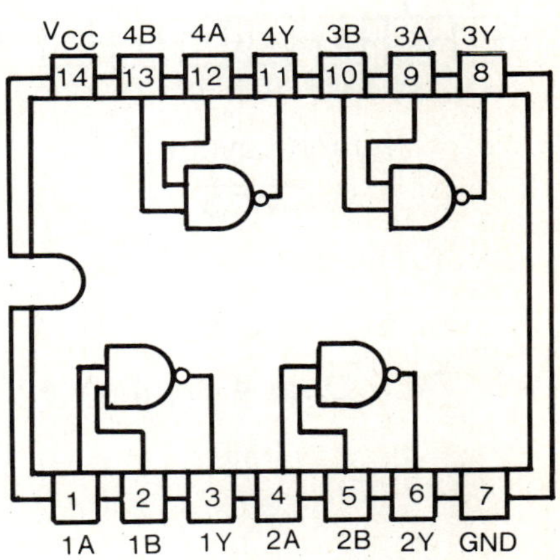

(a) Pin Layout

$Y = \overline{AB}$

(b) Output Functions

Figure D-1. *7400 Quadruple 2-Input NAND Gates*

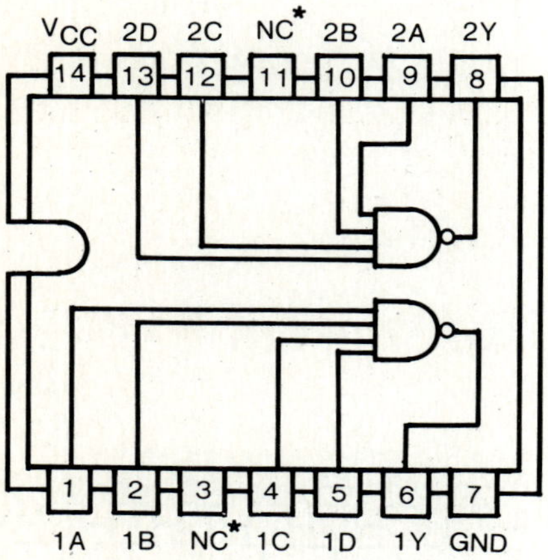

(a) Pin Layout

$Y=\overline{ABCD}$

(b) Output Functions

NC* = NOT USED

Figure D-2. *7420 Dual 4-Input NAND Gates*

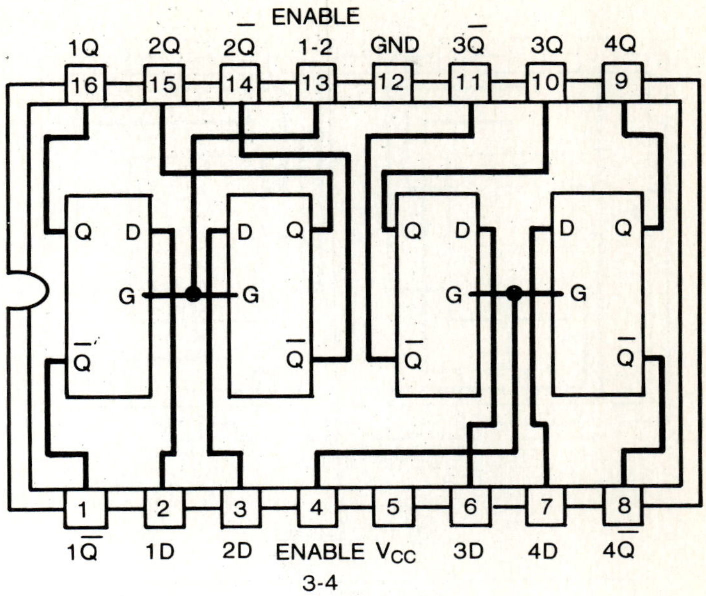

(a) Pin Layout

(EACH LATCH)

INPUTS		OUTPUTS	
D	G	Q	$\overline{Q}$
L	H	L	H
H	H	H	L
X	L	Q_0	$\overline{Q_0}$

(b) Function Table

H= High Level, L=Low Level, X=Irrelevant

Q_0= the level of Q before the high-to-low transition of G

Figure D-3. *7475 4-Bit Bistable Latches*

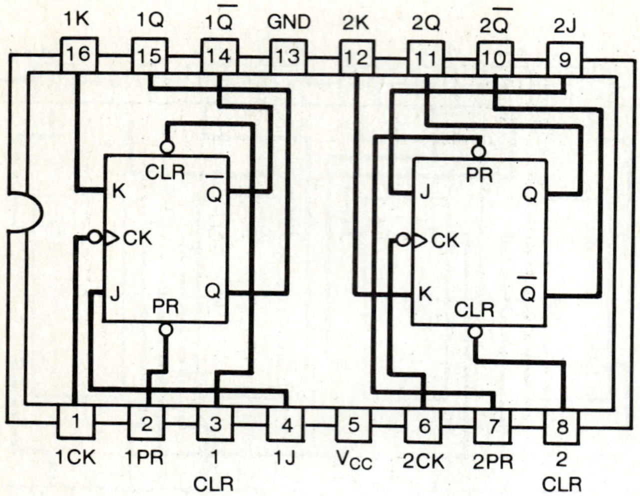

(a) Pin Layout

INPUTS					OUTPUTS	
PRESET	CLEAR	CLOCK	J	K	Q	$\overline{Q}$
L	H	X	X	X	H	L
H	L	X	X	X	L	H
L	L	X	X	X	H★	H★
H	H	⎍	L	L	Q_0	$\overline{Q}_0$
H	H	⎍	H	L	H	L
H	H	⎍	L	H	L	H
H	H	⎍	H	H	TOGGLE	

(b) 7476, 74H76 Function Table

INPUTS					OUTPUTS	
PRESET	CLEAR	CLOCK	J	K	Q	$\overline{Q}$
L	H	X	X	X	H	L
H	L	X	X	X	L	H
L	L	X	X	X	H★	H★
H	H	↓	L	L	Q_0	$\overline{Q}_0$
H	H	↓	H	L	H	L
H	H	↓	L	H	L	H
H	H	↓	H	H	TOGGLE	
H	H	H	X	X	Q_0	$\overline{Q}_0$

(c) 74LS76A Function Table

Figure D-4. *7476 Dual JK Flip-Flops*

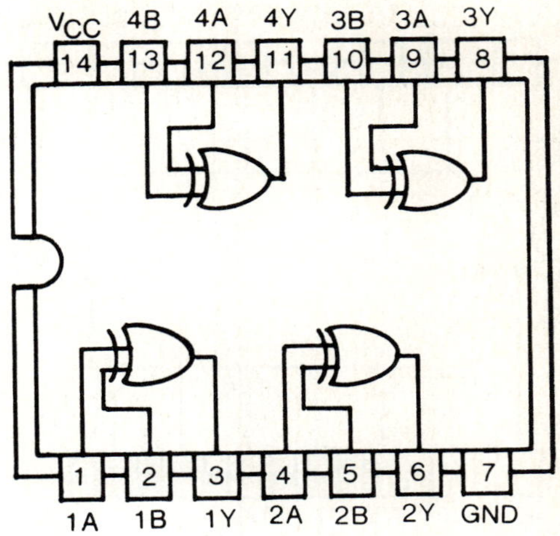

(a) Pin Layout

INPUTS		OUTPUT
A	B	Y
L	L	L
L	H	H
H	L	H
H	H	L

H= HIGH LEVEL, L = LOW LEVEL

(b) Function Table

$$Y = A \oplus B = AB + AB$$

(c) Output Function

Figure D-5. *7486 Quadruple 2-Input Exclusive-OR Gates*

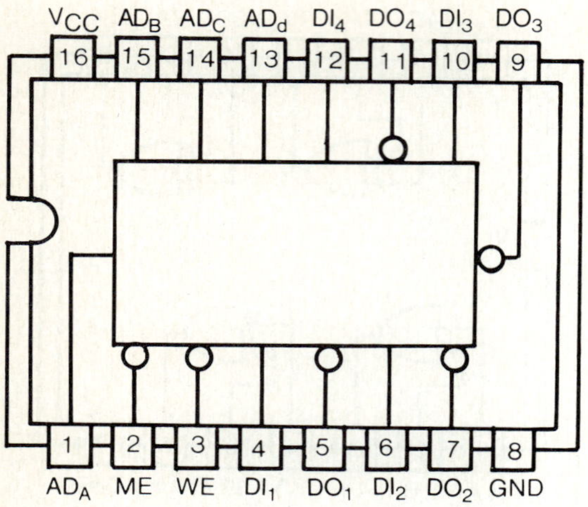

(a) Pin Layout

ME	WE	OPERATION	CONDITION OF OUTPUTS
L	L	WRITE	COMPLEMENT OF DATA INPUTS
L	H	READ	COMPLEMENT OF SELECTED WORD
H	L	INHIBIT STORAGE	COMPLEMENT OF DATA INPUTS
H	H	DO NOTHING	HIGH

(b) Function Table

Figure D-6. *7489 16 4-Bit Memory*

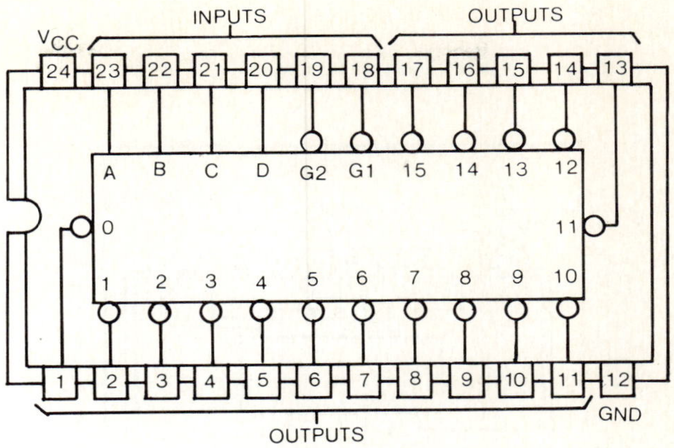

(a) Pin Layout

INPUTS						OUTPUTS															
G1	G2	D	C	B	A	0	1	2	3	4	5	6	7	8	9	10	11	12	13	14	15
L	L	L	L	L	L	L	H	H	H	H	H	H	H	H	H	H	H	H	H	H	H
L	L	L	L	L	H	H	L	H	H	H	H	H	H	H	H	H	H	H	H	H	H
L	L	L	L	H	L	H	H	L	H	H	H	H	H	H	H	H	H	H	H	H	H
L	L	L	L	H	H	H	H	H	L	H	H	H	H	H	H	H	H	H	H	H	H
L	L	L	H	L	L	H	H	H	H	L	H	H	H	H	H	H	H	H	H	H	H
L	L	L	H	L	H	H	H	H	H	H	L	H	H	H	H	H	H	H	H	H	H
L	L	L	H	H	L	H	H	H	H	H	H	L	H	H	H	H	H	H	H	H	H
L	L	L	H	H	H	H	H	H	H	H	H	H	L	H	H	H	H	H	H	H	H
L	L	H	L	L	L	H	H	H	H	H	H	H	H	L	H	H	H	H	H	H	H
L	L	H	L	L	H	H	H	H	H	H	H	H	H	H	L	H	H	H	H	H	H
L	L	H	L	H	L	H	H	H	H	H	H	H	H	H	H	L	H	H	H	H	H
L	L	H	L	H	H	H	H	H	H	H	H	H	H	H	H	H	L	H	H	H	H
L	L	H	H	L	L	H	H	H	H	H	H	H	H	H	H	H	H	L	H	H	H
L	L	H	H	L	H	H	H	H	H	H	H	H	H	H	H	H	H	H	L	H	H
L	L	H	H	H	L	H	H	H	H	H	H	H	H	H	H	H	H	H	H	L	H
L	L	H	H	H	H	H	H	H	H	H	H	H	H	H	H	H	H	H	H	H	L
L	H	X	X	X	X	H	H	H	H	H	H	H	H	H	H	H	H	H	H	H	H
H	L	X	X	X	X	H	H	H	H	H	H	H	H	H	H	H	H	H	H	H	H
H	H	X	X	X	X	H	H	H	H	H	H	H	H	H	H	H	H	H	H	H	H

H=HIGH LEVEL L =LOW LEVEL X=IRRELEVANT/DON'T CARE

(b) Function Table

Figure D-7. *74154 4-Line to 16-Line Decoders/Demultiplexors*

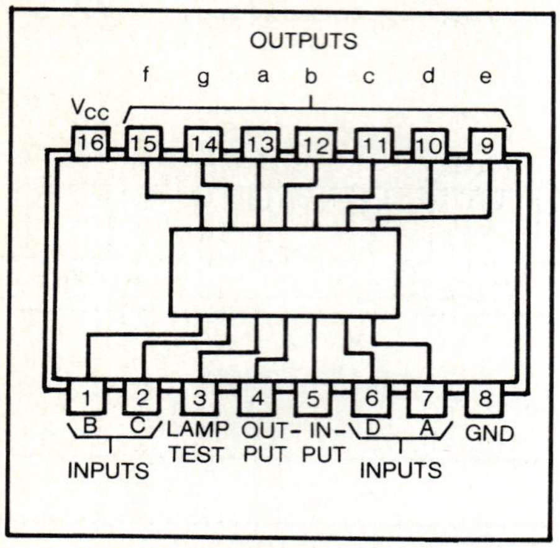

(a) Pin Layout

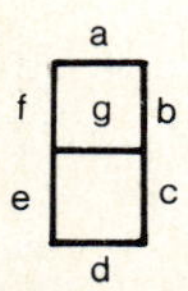

(b) Segment Identification

(c) Numerical Designations and Resultant Displays

DECIMAL OR FUNCTION	INPUTS						OUTPUTS						
	LT	RBI	D	C	B	A	a	b	c	d	e	f	g
0	H	H	L	L	L	L	ON	ON	ON	ON	ON	ON	OFF
1	H	X	L	L	L	H	OFF	ON	ON	OFF	OFF	OFF	OFF
2	H	X	L	L	H	L	ON	ON	OFF	ON	ON	OFF	ON
3	H	X	L	L	H	H	ON	ON	ON	ON	OFF	OFF	ON
4	H	X	L	H	L	L	OFF	ON	ON	OFF	OFF	ON	ON
5	H	X	L	H	L	H	ON	OFF	ON	ON	OFF	ON	ON
6	H	X	L	H	H	L	OFF	OFF	ON	ON	ON	ON	ON
7	H	X	L	H	H	H	ON	ON	ON	OFF	OFF	OFF	OFF
8	H	X	H	L	L	L	ON	ON	ON	ON	ON	ON	ON
9	H	X	H	L	L	H	ON	ON	ON	OFF	OFF	ON	ON
10	H	X	H	L	H	L	OFF	OFF	OFF	ON	ON	OFF	ON
11	H	X	H	L	H	H	OFF	OFF	ON	ON	OFF	OFF	ON
12		X	H	H	L	L	OFF	ON	OFF	OFF	OFF	ON	ON
13	H	X	H	H	L	H	ON	OFF	OFF	ON	OFF	ON	ON
14	H	X	H	H	H	L	OFF	OFF	OFF	ON	ON	ON	ON
15	H	X	H	H	H	H	OFF	OFF	OFF	OFF	OFF	OFF	OFF
LT	L	X	X	X	X	X	ON	ON	ON	ON	ON	ON	ON

(d) Function Table

Figure D-8. *7448 Display/Driver*

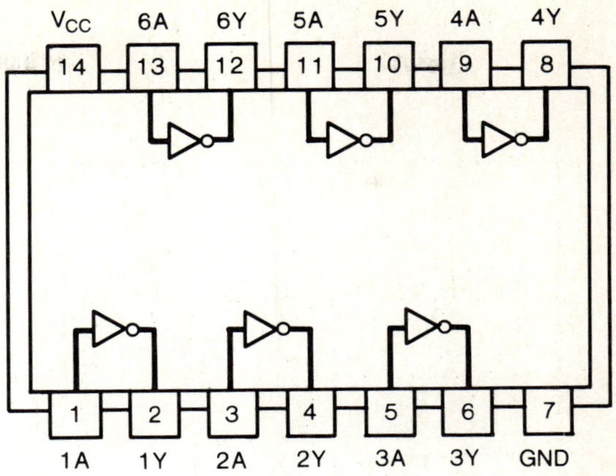

(a) Pin Layout

$Y = \overline{A}$

(b) Output Function

Figure D-9. *7404 HEX Inverter Gates*

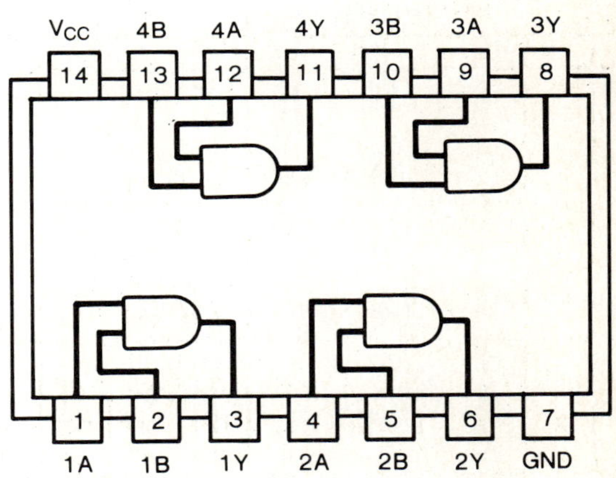

(a) Pin Layout

$Y = AB$

(b) Output Function

Figure D-10. *7408 Quad 2-Input AND Gates*

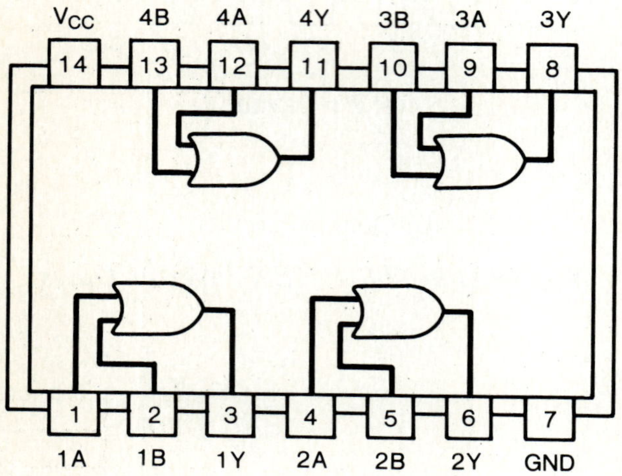

(a) Pin Layout

$$Y = A + B$$

(b) Output Function

Figure D-11. *7432 Quad 2-Input OR Gates*

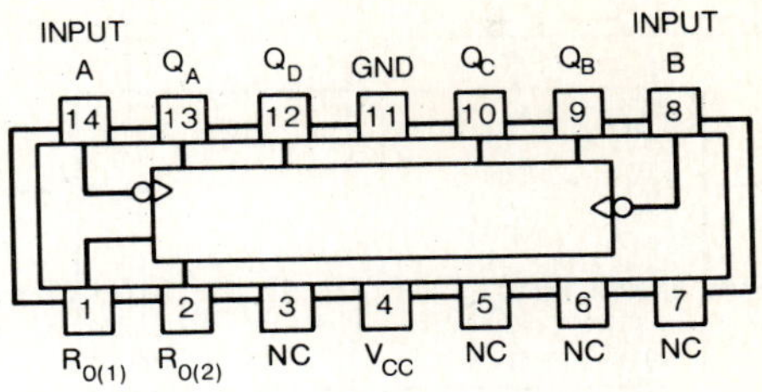

(a) Pin Layout

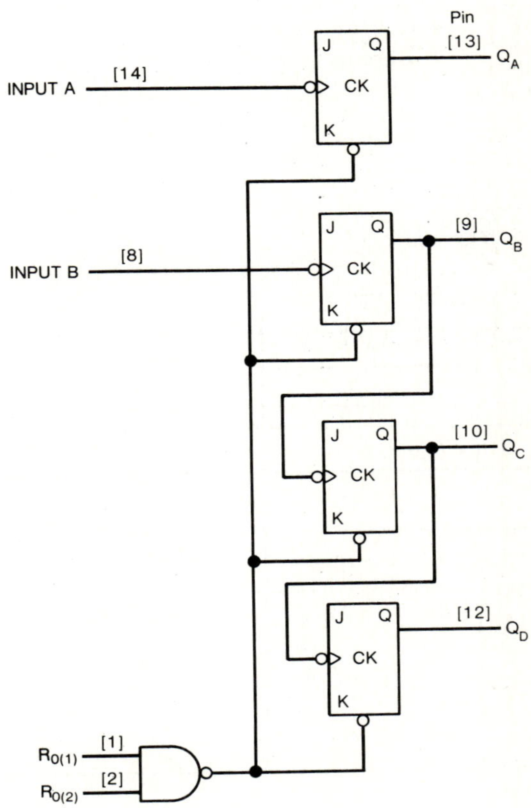

TO USE THE 4-BIT COUNT LENGTH, THE B INPUT IS CONNECTED TO THE Q_A OUTPUT.

(d) Internal Circuit Diagram

RESET INPUTS		OUTPUT			
$R_{0(1)}$	$R_{0(2)}$	Q_D	Q_C	Q_B	Q_Q
H	H	L	L	L	L
L	X	COUNT			
X	L	COUNT			

X=DONT CARE

(b) Reset/Count Function Table

COUNT	OUTPUT			
	Q_D	Q_C	Q_B	Q_A
0	L	L	L	L
1	L	L	L	H
2	L	L	H	L
3	L	L	H	H
4	L	H	L	L
5	L	H	L	H
6	L	H	H	L
7	L	H	H	H
8	H	L	L	L
9	H	L	L	H
10	H	L	H	L
11	H	L	H	H
12	H	H	L	L
13	H	H	L	H
14	H	H	H	L
15	H	H	H	H

(c) Count Truth Table

Figure D-12. *7493 4-Bit Binary Counter*

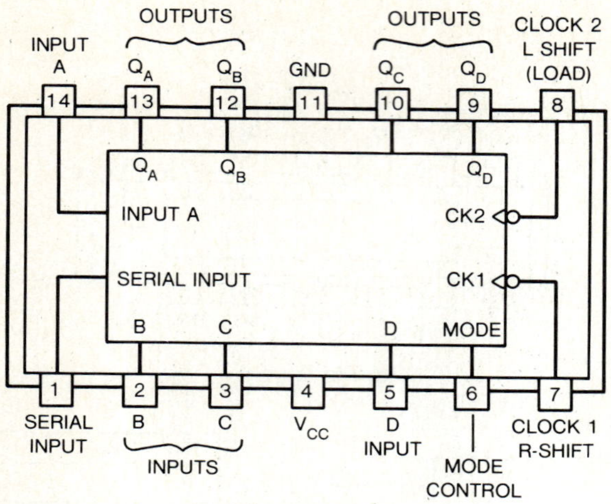

(a) Pin Layout

OPERATING MODE	INPUTS									OUTPUTS			
	MODE CONTROL	CLOCKS		SERIAL	PARALLEL								
		2 (L)	1 (R)		A	B	C	D		Q_A	Q_B	Q_C	Q_D
PARALLEL LOAD	H	↓	X	X	A	B	C	D		A	B	C	D
SHIFT LEFT*	H	↓	X	X	Q_B	Q_C	Q_D	D		q_B	q_C	q_D	D
SHIFT RIGHT	L	X	↓	S	X	X	X	X		S	q_A	q_B	q_C
MODE CHANGE	↑	L	L	X	X	X	X	X		NO CHANGE			
	↓	L	L	X	X	X	X	X					
	↓	L	H	X	X	X	X	X					
	↑	H	L	X	X	X	X	X					
	↑	H	H	X	X	X	X	X					

**SHIFTING LEFT REQUIRES EXTERNAL CONNECTIONS OF PIN 12 TO 14, 10 TO 2, 9 TO 3. SERIAL LEFT SHIFT INPUT IS ENTERED AT INPUT D.*

(b) Function Table

Figure D-13. *7495 Parallel Access Shift Register*

Bibliography

There are quite a few books which serve as good reference to logic and computer design. The reference sources that the authors have consulted in writing the book are listed below:

Logic Design

1. Blakeslee, T. R., *Digital Design with Standard MSI and LSI* (2d Ed.). New York: John Wiley & Sons, 1978.
2. *Digital System Laboratory Manual.* Electrical Engineering Department, Illinois Institute of Technology, 1980.
3. *Digital Techniques, Student Workbook.* Heathkit Learning Publication, 1978.
4. Greenfield, J. D., *Practical Digital Design Using ICs.* New York: John Wiley & Sons, 1979.
5. *Harris Bipolar and CMOS Digital Databook*, Vol. 2. Harris Corporation, 1981.
6. Heiserman, P. L., *Handbooks of Digital IC Applications.* Englewood Cliffs, N.J.: Prentice-Hall, 1980.
7. Leach, D. P., *Experiments in Digital Principles* (2d Ed.). New York: McGraw-Hill, 1981.

8. Rony, P. R. and Larsen, D. G., *The Bugbook.* E&L Instruments, Inc., 1974.

9. *The TTL Data Book for Design Engineers* (2d Ed.). Texas Instruments, 1976.

10. Tomek, I., *Introduction to Computer Organization Workbook.* Potomac, Maryland: Computer Science Press, 1981.

11. Wakerly, J. F., *Logic Design Projects Using Standard Integrated Circuits.* New York: John Wiley & Sons, 1976.

12. Williams, G. E., *Digital Technology.* Chicago: SRA, 1977.

13. ____, *Digital Technology Laboratory Manual.* Chicago: SRA, 1977.

Computer Design

1. Crane, J., *Laboratory Experiments for Microprocessor Systems.* Englewood Cliffs, N.J.: Prentice-Hall, 1980.

2. Fletcher, W. I., *An Engineering Approach to Digital Design.* Englewood Cliffs, N.J.: Prentice-Hall, 1980.

3. Hill, F. J. and Peterson, G. R., *Digital Systems: Hardware Organization and Design.* New York: John Wiley & Sons, 1978.

4. Kraft, G. D. and Toy, W. N., *Mini/Microcomputer Hardware Design.* Englewood Cliffs, N.J.: Prentice-Hall, 1979.

5. Mano, M. M., *Computer System Architecture* (2d Ed.). Englewood Cliffs, N.J.: Prentice-Hall, 1983.

6. ____, *Digital Logic and Computer Design.* Englewood Cliffs, N.J.: Prentice-Hall, 1979.

7. Winkel, D. and Prosser, F., *The Art of Digital Design - An Introduction to Top-Down Design.* Englewood Cliffs, N.J.: Prentice-Hall, 1980.

VLSI Systems

1. Lee., D. T., et al., "An ON-CHIP Compare/Steer Bubble Sorter," *IEEE Transactions on Computers*, Vol. C-30, No. 6, June, 1981, pp. 396-405.

2. Mead, C. and Conway, L., *Introduction to VLSI Systems.* Reading, Massachusetts: Addison-Wesley, 1980.

Computer Communications

1. Liu, M. T. and Reames, C. C., "The Design of the Distributed Loop Computer Network," *Proc. International Computer Symposium*, Taipei, Taiwan, 1975, Vol. I, pp. 273-284.

2. Tanenbaum, A. S., *Computer Networks*. Englewood Cliffs, N.J.: Prentice-Hall, 1981.

3. Teng, A. Y., *Protocol Constructions for Computer Networks*, Ph.D. Thesis, Department of Computer and Information Sciences, Ohio State University, 1980.

4. Teng, A. Y., Yao, J., Gopinath, B., and Sabnani, K., "A File Transfer System for Scheduling File Transfers in BLN," *Proc. INFOCOM 83*, San Diego, California, April 20, 1983, pp. 279-287.

Index

H

I

J

K

L

M

N

O

P